David G. Martinez

P. Michigan XIX

Baptized for Our Sakes:
A Leather Trisagion from Egypt
(P. Mich. 799)

# Beiträge zur Altertumskunde

Herausgegeben von
Michael Erler, Ernst Heitsch, Ludwig Koenen,
Reinhold Merkelbach, Clemens Zintzen

Band 120

B. G. Teubner Stuttgart und Leipzig

# P. Michigan XIX

## Baptized for Our Sakes: A Leather Trisagion from Egypt

*(P. Mich. 799)*

Edition and Commentary
by
David G. Martinez

B. G. Teubner Stuttgart und Leipzig 1999

Die Deutsche Bibliothek – CIP-Einheitsaufnahme

**Martinez, David G.:**
Baptized for our sakes : a leather trisagion from Egypt (P. Mich. 799) ;
edition and commentary / by David G. Martinez –
Stuttgart ; Leipzig : Teubner, 1999
(Beiträge zur Altertumskunde ; Bd. 120) (P. Michigan ; 19)
ISBN 3-519-07669-1



Printed in Germany
Druck und Bindung: Druckhaus „Thomas Müntzer“ GmbH, 99947 Bad Langensalza

To my mother
and the memory of my father

## TABLE OF CONTENTS

# PREFACE

In preparing this edition, I received generous assistance from a National Endowment for the Humanities Fellowship with a matching University Research Institute grant from the University of Texas at Austin and two Summer Research Awards from the same university. I acknowledge my gratitude for this support and also for permission to publish this remarkable document, granted by the Hatcher Graduate Library of the University of Michigan, Ann Arbor. I have benefited greatly from the correspondence of Rev. Fr. Justin of St. Catherine's Monastery, Mt. Sinai, and Rev. Fr. Haralampos of the Holy Transfiguration Monastery, Brookline, MA, who freely made available their expertise in Eastern Orthodox liturgy and theology. I also thank my Texas colleague David Armstrong for contacting them in my behalf and for his own valuable suggestions.

I made substantial revisions to the text and finished checking references during my fellowship year at the Center for Hellenic Studies in Washington, D.C., where, in addition to the Center's resources, I also had access to the splendid Byzantine library at Dumbarton Oaks. Those two institutions have done so much to advance ancient and late-antique Greek studies in this country, that I am honored to have this monograph associated with them. I am also pleased for the volume to bear the name of B.G. Teubner, and I appreciate the patience of the editors of that firm in bringing it to light.

I presented various aspects of this text as papers to the American Society of Papyrologists, the Catholic University of America, the Center for Hellenic Studies, and Dumbarton Oaks, and benefited from the criticisms of those audiences, particularly L. MacCoull, J.

Price, and A. Alexakis. I also thank T. Gagos, F. T. Gignac, P.A. Heilporn, R. Hübner, S. Larson, M. Marcovitch, G. Quispel (†), A. Rigsby, R.K. Ritner, G. Schwendner, L. M. White, and D. Wilson for their help and advice on many points. In the production of the camera-ready copy I received invaluable assistance from Jeffrey Fish, whose expertise with computers and keen critical insight improved this work on a number of fronts. In addition, I am grateful to my parents-in-law, Thomas and Irma Longley, to my aunt, Antonia Cabrera, and above all to my wife Meredith for their support of a more personal but equally important nature.

As I wrote this monograph two were looking over my shoulder. The first is Ludwig Koenen, who gave me constant help with this document and whose own ground-breaking work on the Trisagion forms the basis for my approach to this type of hymn in general and this text in particular. The other is my late father, Rev. Angel Martinez, whose character and long career in the ministry furnished a kind of spiritual incentive for my work. I dedicate this volume to his memory and also to his surviving wife and my mother, Robbie Martinez.

Austin, Texas

February 1999

## Editorial Sigla

The editorial *sigla* employed in this study are those commonly used in editions of papyri.

| | |
|---|---|
| [ ] | lacuna in text |
| ( ) | resolution of an abbreviation or symbol |
| { } | superfluous letter or letters |
| ` ´ | additions above the line |
| ⟦ ⟧ | deletion in the original |
| < > | omission in the original |

A dot placed beneath a letter indicates that the letter is doubtful. Within square brackets dots indicate the estimated number of missing letters.

# I. INTRODUCTION[1]

The Michigan Papyri do not contain a large number of Christian texts. Most of the material which fall under this rubric are fragments from Biblical and Patristic writings. This well-preserved leather document[2] (c. 7th cent.) is one of the few hymnic pieces to surface in that collection. Its thirty-one lines of text comprise three separate hymns which all have in common the well-known Trisagion refrain, ἅγιος ὁ θεός, ἅγιος ἰςχυρός, ἅγιος ἀθάνατος, ἐλέηςον ἡμᾶς. The first hymn is an acrostic recounting of the life and passion of Jesus with three repetitions of the Trisagion interposed upon it (1-39; dipl. 2-14). The second consists of an introduction, narrating Isaiah's vision of the seraphim and its significance, followed by a hymnic arrangement of Biblical allusions, with each refrain of the Trisagion having a three-fold ἅγιος instead of one (40-54; dipl. 15-22). The third hymn is a series of LXX quotes interlaced with the standard Trisagion (55-65; dipl. 22-28),[3] followed by a sextuplet which concludes the entire document (66-70; dipl. 28-31).

## Language and Rhythm

*P. Mich.* 799 is a virtual compendium of the numerous vowel interchanges, consonant variations, omissions, and additions which were rampant in papyri of this era. Our hymn, however, supersedes

---

[1]The references to *P. Mich.* 799 in this introduction are to the edited/metrical text (see below pp. 38ff.) unless otherwise noted.

[2]For the production and use of leather for a writing surface, cf. E.G. Turner, *Greek Papyri* (Oxford 1980) 8f. with n. 40 (p. 182); R.J. Forbes, *Studies in Ancient Technology* V (Leiden 1966[2]) 63ff.; R. Reed, *Ancient Skins, Parchments and Leathers* (London 1972) passim, esp. 118-23. ♦

[3]The integrating of scripture into hymn owes much to the style of Revelation, where scripture is never directly quoted with one of the traditional formulae, such as γέγραπται γάρ, but is interlaced with hymnic motifs (cf. N. Turner, *A Grammar of New Testament Greek*, vol. 4, *Style*, [Edinburgh 1976] 145).

even most Byzantine-period documents in the sheer volume and variety of its irregularities. In Roman and Byzantine documentary texts the most common interchanges are (in descending order) ι/ει, ε/αι, and υ/οι.[4] In the Michigan Trisagion the first occurs eight times, the second twice, the third only once. By far the most frequent interchange is ο ⟩ ω, occurring 20 times, whereas ω ⟩ ο occurs only once.[5] Our text also shows a number of phonological characteristics which betray bilingual interference,[6] such as the frequent ε/η exchange and the less frequent exchanges of ου/ω(ι), ο/ε, η/υ, οι/ε, and οι/η. On the particulars see the diplomatic transcript, phonological appendix and ad locc.

The "metrical" system on which the first two hymns are based is accentual. Within the various hymnic units the cola have the same number of major stress accents (with the exception of final cola in certain groups, which may act as clausulae and have additional stresses[7]). By "major" we mean the accents borne by major words; those of articles, prepositions, many pronouns,[8] and some other *kleine Wörter* do not count. This rhythmic structure owes its origins to Semitic models of poetry[9] rather than classical Greek meters. End-

---

[4]Gignac I 192, 197. These exchanges and ο/ω will not be documented in the commentary.

[5]Although ο/ω is extremely common as a general phenomenon in papyri of all periods (esp. before nasals, ρ, and ϲ; Mayser/Schmoll I 1.73-76; Teodorsson 151-59, 233f.; Gignac I 275-77), its prolific use in individual documents is idiosyncratic. Cf. the papyri of Apollonios, brother of Ptolemaios, (Mayser/Schmoll I 1.73, 76) and *P. Oxy.* XVI 1880 (AD 427) with its 9 examples (Gignac I 277). More recently, see *Suppl. Mag.* II 58 (IV-V AD); MPER N.S. XVII 37 (VII-VIII AD).

[6]In general see Gignac I 46ff.; Teodorsson 58 n. 187.

[7]In our text lines 44 and 52, and cf. Koenen 33. On accentual verse in general see Koenen 32-34; idem, *BASP* 22 (1985) 173-78; Maas/Trypanis 511ff.; Römer p. 70f.

[8]For the ambiguous position of pronouns in this rhythmic scheme, see below p. 54 on 14-15 and p. 57 on 20-21, and in general Maas/Trypanis p. 512.

[9]It is, however, difficult to determine precisely which of the Semitic traditions or what combination of them supplied the influence; cf. Koenen 34 n. 6;

rhyme, which sometimes characterizes this kind of verse, does not occur in our text outside the Trisagion cola themselves.[10]

This type of poetry is not only to be distinguished from quantitative Greek verse but also from the isotonic/isosyllabic Byzantine church poetry which developed from the classical tradition. An example of the latter is Maas *FK* I 1.1-3:[11]

| | Syllables | Stresses |
|---|---|---|
| ἡ ἀϲώματοϲ φύϲιϲ τῶν χερουβὶμ | 11 | 3 ($xx\underline{x}xx\underline{x}xxxx\underline{x}$) |
| ἀϲιγήτοιϲ ϲε ὕμνοιϲ δοξολογεῖ | 11 | 3 ($xx\underline{x}xx\underline{x}xxxx\underline{x}$) |
| ἑξαπτέρυγα ζῷα, τὰ ϲεραφίμ, | 11 | 3 ($xx\underline{x}xx\underline{x}xxxx\underline{x}$) |
| ταῖϲ ἀπαύϲτοιϲ φωναῖϲ | | |
| ϲε ὑπερυψοί· | 11 | 3 ($xx\underline{x}xx\underline{x}xxxx\underline{x}$) |
| τῶν ἀγγέλων δὲ πᾶϲαι αἱ ϲτρατιαὶ | 11 | 3 ($xx\underline{x}xx\underline{x}xxxx\underline{x}$) |
| τριϲαγίοιϲ ϲε ᾄϲμαϲιν εὐφημεῖ· | 11 | 3 ($xx\underline{x}xx\underline{x}xxxx\underline{x}$) |

---

idem, *BASP* 22 [1985] 174 n. 8. Principles of Semitic rhythmization not only provided a basis for certain types of accentual poetry, but also influenced the style of theological prose, both pagan (e.g., the "Dream of Nectanebos"; L. Koenen, *BASP* 22 [1985] 171-94, esp. 173-76) and Christian (e.g., Didymos the Blind; see the ed. of *de Trinitate* 2.1-7 by I. Seiler (Beiträge zur klassischen Philologie 52, Meisenheim 1975) viii-xiii; L. Koenen, *APF* 17 (1960) 91-99.

[10]On the Trisagion cola see below p. 9 n. 23. For rhyme in accentual Christian poetry, see A. Dihle, *Hermes* 82 (1954) 198; W. Meyer, *Abhandlungen der Akademie der Wissenschaften München* 17 (1886) 377ff.; Norden, *Ag. Th.* 262f.

[11]So also Trypanis, "Three Hymns," but with the third stress on the antepenultimate syllable rather than the ultimate; cf. the metrical observations on the poems by A. Dihle, *BZ* 69 (1976) 1-5. These principles work differently in the poetry of Romanos, where the syllable count and major stress points in the lines of each stanza conform to those of the first stanza ("outer correspondence"; see Maas/Trypanis 511). Cf. also *P.Köln* IV 173, with individual cola having different numbers of syllables, but the total syllable count of each strophe being the same (on which see Römer pp. 68-75). It has been held that this type of poetry was based on Syriac models, particularly on the hymnic form called the *madrasa*, as perfected by Epraim Syrus (see below p. 27). A. Dihle, however, has shown that it rather originated from the correlation between ictus and accent as developed in later Greek poetry (*Hermes* 82 [1954] 182-199, esp. 190ff., and for its distinction from the more Semitic-influenced material, 198f.).

Compare with this the more Semitic style of the Greek "synagogue prayer" from *constitutiones apostolorum* 8.15.7:[12]

| | Syllables | Stresses |
|---|---|---|
| ὁ θεὸς ὁ παντοκράτωρ | 8 | 2 |
| ὁ ἀληθινὸς καὶ ἀσύγκριτος | 9 | 2 |
| ὁ πανταχοῦ ὢν | 5 | 2 |
| καὶ τοῖς πᾶσιν παρὼν | 6 | 2 |
| καὶ ἐν οὐδενὶ ὡς ἐνόντι ὑπάρχων, | 12 | 3 |
| κτλ. | | |

The haphazard syllable count conveys a more prosaic feel. With this we may compare two sections of our text: the preface to the second hymn (40-45), and the concluding sextuplet (66-70), with their regular (or nearly regular) accentual pattern but random syllable count (the third hymn [55-65], consisting entirely of LXX quotes, is irregular in both respects; see ad loc.). The second hymn (46-54), however, is isosyllabic, with a consistent pattern of seven syllables per colon (except in the clausula and the Trisagion stanzas). Hymn 1

[12]As cited by A. Dihle, *Hermes* 82 (1954) 198. For these prayers (II/III AD) see the introduction and bibliography by D.A. Fiensy and the Eng. trans. with notes by D.R. Darnell in J.H. Charlesworth (ed.), *The Old Testament Pseudepigrapha* 2 (Garden City, NY, 1985) 671ff. (and cf. 694f. for this passage). We see a similar rhythmic scheme in the opening lines of the LXX version of the "Song of Moses" (Ex. 15.1-2):

| | Syllables | Stresses |
|---|---|---|
| Ἄισωμεν τῷ κυρίῳ, ἐνδόξως γὰρ δεδόξασται | 15 | 4 |
| ἵππον καὶ ἀναβάτην ἔρριψεν εἰς θάλασσαν | 14 | 4 |
| βοηθὸς καὶ σκεπαστὴς ἐγένετό μοι εἰς σωτηρίαν· | 17 | 4 |
| οὗτός μου θεός, καὶ δοξάσω αὐτόν, | 11 | 4 |
| θεὸς τοῦ πατρός μου, καὶ ὑψώσω αὐτόν. | 12 | 4 |

For other examples of this style, see the "angelic" Trisagion (below p. 7) and the Trisagion in the liturgy of the Coptic Jacobites (below p. 19). The Manichaean Parthian hymns were also apparently composed after this manner (M. Boyce, *A Reader in Manichaean Middle Persian and Parthian* [Acta Iranica 9, Leiden 1975] p. 13).

(1-39) displays an almost consistent pattern of stresses and, with few exceptions, varies between six or seven syllables per colon, which is neither strictly isosyllabic nor completely random in the same way as the above-cited synagogue prayer and the parallel passages of our document. None of the poems of our text show isotonic principles. Thus, our hymns 1 and 2 seem to occupy a middle ground between the more stylized Byzantine hymns and the purely accentual Semitic-influenced material. This should not surprise us. Although the two types of poetry have different origins, they share accentual verse and developed within the religious traditions of eastern Christianity. It would thus be surprising if the two did not evince mutual influence and encroach on each other's territory. An examination of hymns from the Byzantine and Oriental[13] traditions in fact reveals many gradations between the highly structured Byzantine hymn, as above represented in ἡ ἀcώματοc φύcιc, and the purely accentual Semitic-influenced poem, such as the selection from *const. ap.* 8.[14]

[13]By the terms "Byzantine" and "Oriental" I mean the ecclesiastic network and theological tradition (also called "Eastern Orthodox") which accepted the Calcedonian Christological definitions as opposed to the churches and theologians who did not, the latter being specifically the Armenian, Egyptian/Coptic, Ethiopian, and Syrian Orthodox communions (see *ODCC*[3] s.vv. "Oriental Orthodox Churches" [p. 1193] and "Orthodox Church" [p. 1197]). Cf. the distinction made by Ephraim, the sixth-cent. Patriarch of Antioch, between οἱ μὲν τὴν ἀνατολὴν οἰκοῦντεc, who sing the Trisagion to Jesus, and οἱ δὲ Βυζάντιόν τε καὶ τὴν ἑcπερίαν νεμόμενοι, who sing it to the Trinity (apud Photios, *biblioth.* 228, cited fully below p. 17f.). Potential for confusion arises, when papyrologists speak of papyri dating after the mid-fourth cent. AD, including theological and liturgical texts, as Byzantine, although their religious tradition, being from Egypt, is often rather "Oriental." When I so use "Byzantine" as a chronological label, I append a word such as "period" unless the context leaves no room for ambiguity.

[14]Rhythmic poetry exerted its influence on Christian literature from its beginnings. NT 1Tim. 3.16, for example, evinces a regular pattern of six cola with two stresses per cola (cf. Koenen 33 n. 4; Norden, *Ag. Th.* 254ff.). Cf. also Eph. 5.14:

| | **Syllables** | **Stresses** |
|---|---|---|
| ἔγειρε, ὁ καθεύδων, | 7 | 2 |
| καὶ ἀνάcτα ἐκ τῶν νεκρῶν, | 8 | 2 |
| καὶ ἐπιφαύcει coι ὁ Χριcτόc. | 9 | 2 |

## The Trisagion—Structure, Epithets, and Origins

For the church in both the East and West, the supreme expressions of worship were those which imitate that of the heavenly realm.[15] For this reason Biblical texts which describe the adoration

---

On this text and for theories with regard to its liturgical use in the early church, see C.F.D. Moule, *The Birth of the New Testament* (San Francisco, $1982^3$) 34f. Attempts, however, to establish regular rhythmic patterns for other "hymnic" passages such as Jn. 1.1-14, Col. 1.15-20, and esp. Phil. 2.6-11 have not succeeded (cf. Moule, op. cit. 35ff.; Norden, *Ag. Th.* 250-63; G. Delling in *TDNT* VIII 500ff. s.v. ὕμνος; R.P. Martin, *Carmen Christi* [Society for New Testament Studies Monograph Series 4, Cambridge 1967] 17-23). Their rhythms parallel more closely the irregular patterns of Hebrew poetry, which scholars have found just as difficult to describe in terms of consistent principles. There is, in fact, little agreement among Hebraists whether parallelism, accentuation, isosyllabism, or other criteria form the basis of Hebrew meter or even if Hebrew verse can be said to have meter at all (cf. in general D.K. Stuart, *Studies in Early Hebrew Meter* [Harvard Semitic Monograph Series 13; Missoula 1976] 1-20; M. O'Connor, *Hebrew Verse Structure* [Winona Lake, 1980] 29-67; and most recently S.E. Gillingham, *The Poems and Psalms of the Hebrew Bible* [Oxford 1994] 44-68). The similarities between the hymnic sections of the New Testament and Hebrew verse are not borne of any conscious imitation by the authors of the former, but is rather the natural outgrowth of a religious community imbued in the language of the LXX. This is not surprising, since the portions of the LXX to which Christians from the New Testament onward have most often appealed for the proof and defense of their faith are the poetry of the prophetic works and the Psalms (cf. H.M. Shires, *Finding the Old Testament in the New* [Philadephia 1974] 66-76, esp. 71f.; B. Lindars, *New Testament Apologetic* [Philadelphia 1961] 272ff.).

[15]To my knowledge, the earliest expression of this sentiment in Christian literature is 1Clem. 34.5-7. After citing LXX Dan. 7.10 and Is. 6.3, which both describe angelic worship, Clement exhorts: καὶ ἡμεῖϲ οὖν ἐν ὁμονοίᾳ ἐπὶ τὸ αὐτὸ ϲυναχθέντεϲ τῇ ϲυνειδήϲει, ὡϲ ἐξ ἑνὸϲ ϲτόματοϲ βοήϲωμεν πρὸϲ αὐτὸν ἐκτενῶϲ εἰϲ τὸ μετόχουϲ ἡμᾶϲ γενέϲθαι τῶν μεγάλων καὶ ἐνδόξων ἐπαγγελιῶν αὐτοῦ. Cf. Chrys. *hom. in Is.* 6.1, 1.1.31ff., Dumortier SC 277 (Migne PG 56.97): ἄνω ϲτρατιαὶ δοξολογοῦϲιν ἀγγέλων· κάτω ἐν ἐκκληϲίαιϲ χοροϲτατοῦντεϲ ἄνθρωποι τὴν αὐτὴν ἐκείνοιϲ ἐκμιμοῦνται δοξολογίαν. ἄνω τὰ ϲεραφὶμ τὸν τριϲάγιον ὕμνον ἀναβοᾷ· κάτω τὸν αὐτὸν ἡ τῶν ἀνθρώπων ἀναπέμπει πληθύϲ (cf. Brightman 479 n. 20). Fr. Justin points out to me that Christian authors found Biblical grounds for this belief in NT Hebrews, which portrays the earthly Levitical cult in Platonic/Philonic cate-

of God and/or Christ by angels or other divine beings, such as the book of Revelation and various Old Testament prophetic writings, set a standard for liturgical conception and practice. Perhaps the most famous of these is the passage on which the Trisagion is based, Isaiah's visionary account of the seraphim surrounding the throne of Yahweh and exclaiming to one another (6.3):[16]

| | Syllables | Stresses |
|---|---|---|
| ἅγιος ἅγιος ἅγιος κύριος Σαβαωθ, | 15 | 5 |
| πλήρης πᾶσα ἡ γῆ τῆς δόξης αὐτοῦ | 11 | 5[17] |

This seraphic worship focuses on divine transcendence and otherness. The angelic beings dare not look upon Yahweh, but hide their faces with their wings and acclaim him in the third person (*Er-Stil*). Most Greek liturgical traditions, however, slightly expand the second colon and modify it to *Du-Stil*: πλήρης ὁ οὐρανὸς καὶ ἡ γῆ τῆς δόξης σου.[18] In addition, the *interpretatio christiana* of the eastern churches united Isaiah's angelic Trisagion with that of the New Testament Revelation 4.8, where the cherubim sing a slightly different version of the hymn to the enthroned Father and Son.[19] Thus various sources attribute it to both seraphim and cherubim or to the cherubim

---

gories as a reflection of the heavenly. E.g., 8.5, which describes the Jews as οἵτινες ὑποδείγματι καὶ σκιᾷ λατρεύουσιν τῶν ἐπουρανίων, and see Chrysostom's famous comment on this passage, οὐρανία ἐστίν ἡ ἐκκλησία, καὶ οὐδέν ἐστιν ἄλλο ἢ οὐρανός (*hom. in Hebr.* 14.2, Migne PG 63.112). For further material and discussion see J. Daniélou, *The Bible and the Liturgy* (Notre Dame 1956; repr. Ann Arbor 1979) 135; Werner 19-29.

[16]ἕτερος πρὸς τὸν ἕτερον suggests a liturgical framework. Cf. Paul's instructions for worship in Eph. 5.19: (ὑμεῖς) λαλοῦντες ἑαυτοῖς ἐν ψαλμοῖς καὶ ὕμνοις καὶ ᾠδαῖς πνευματικαῖς (similarly Col. 3.16).

[17]The rhythmic scheme is of the same type as the "synagogue prayer"; see above p. 4.

[18]e.g., *lit. Jo. Chrys.*, Brightman 385.10; see Brock 24f. The early Syrian rite preserved in the *Apostolic Constitutions* retains the αὐτοῦ of the Biblical text after δόξης (Brightman 19.1).

[19]See Brock 26. On Revelation's version see below.

alone.[20] In the mind of Byzantine and Oriental Christianity both Isaiah and the seer of the Apocalypse contemplated the same timeless vision of perpetual heavenly worship.

It is difficult to exaggerate the significance which this angelic hymn held for the church, especially in the East. Among Christians of the Greek and Oriental Orthodox communions it approximated the esteem of the *shema* in Judaism, not only as a liturgical proclamation of religious devotion, but also as a magical formula of power.[21] It is not surprising that so great an affirmation of divine nature generated a second form. The three ἅγιοι of this "angelic" Trisagion were expanded by the inclusion of epithets into the "ecclesiastic" Trisagion:[22]

---

[20]Both: Maas *FK* I 1.1ff., 2.6, 5.19; cherubim only: *lit. Mk.*, Brightman 122.18ff.; *lit. Chrys.* idem 377.9ff. The celestial chorus sometimes included not only the cherubim and seraphim, but other New Testament angelic ranks as well: cὲ προcκυνοῦcιν ἀναρίθμητοι cτρατιαὶ ἀγγέλων ἀρχαγγέλων θρόνων κυριοτήτων ἀρχῶν ἐξουcιῶν δυνάμεων cτρατιῶν αἰωνίων, τὰ χερουβὶμ καὶ τὰ ἑξαπτέρυγα cεραφὶμ --- λέγοντα ἅμα χιλίαιc χιλιάcιν ἀρχαγγέλων καὶ μυρίαιc μυριάcιν ἀγγέλων ἀκαταπαύcτωc καὶ ἀcιγήτωc βοῶcαιc, "ἅγιοc ἅγιοc ἅγιοc" κτλ. (*lit. Clem.*, Brightman 18.25ff.). Cf. *lit. Jac.*, Brightman 50.16ff.; *lit. Mk.*, idem 131.21ff.; *lit. Copt. Jac.*, idem 175.22-176.3; *lit. Chrys.*, idem 313.8ff.; Maas *FK* I 2.5ff.; Schermann 54-55; Brock 26f. For these various angelic orders, see Eph. 1.21; Col. 1.16; Michl in *RAC* V 175-82, esp. with regard to the Trisagion 178f.; Daniel/Maltomini, *Suppl. Mag.* I p. 94 (on 32.5-6).

[21]The hymn (or variations on it) appears frequently in Christian protective /exorcistic amulets from the fifth century onward: P. Heid. G.1101.1, 6 with n. p. 158, where Maltomini cites extensive parallels (*ZPE* 48 [1982] 149-70 [*Suppl. Mag.* I 32]); *PGM* P13.7 (*Ancient Christian Magic*, edd. M. Meyer and R. Smith [San Francisco 1994] #10, pp. 35f.); ibid. 18.1 (*An. Christ. Mag.* #13, pp. 38f.); *P.Köln* IV 171.9 (with n. ad loc.); Daniel/Maltomini *Suppl. Mag.* I 25.5ff., 29.15ff., 36.19 (Lat.). For further material and discussion, see J. Russell in *Byzantine Magic* (Washington, D.C. 1995) ed. H. Maguire, 39 with n. 13; Th. Klauser in *RAC* I 230 s.v. Akklamation; E. Peterson, *Heis Theos* (Forschungen zur Religion und Literatur des Alten und Neuen Testaments, Neue Folge 24) Göttingen 1926, 233f., 325. *PGM* O.3, which preserves a farced ecclesiastic Trisagion, is not magical (see Koenen 39).

[22]Ancient and modern sources often use simply "Trisagion" for both (cf. the above [n. 15] cited Chrys. *hom. in Is. 6.1*, I 1.31ff.), which has caused some confusion (see Quecke, *Stundengebet*, 24 with n. 52; so Lat. *tersanctus*; see *ODCC*[3]

ἅγιοϲ ὁ θεόϲ, ἅγιοϲ ἰϲχυρόϲ, ἅγιοϲ ἀθάνατοϲ,[23] and in the spirit of Isaiah, who expressed inadequacy before the divine holiness (6.5), ἐλέηϲον ἡμᾶϲ was added.[24]

---

1591 s.v.). I will hereafter distinguish between the two with the terms "angelic" and "ecclesiastic." "Trisagion" without a qualifier refers to the ecclesiastic hymn. Various traditions relate different narratives of its supernatural origins (in general see Brock 28f.). For the account that it was revealed to a child by the angels during an earthquake in Constantinople, see Brightman 530f. n. 2; Hanssens p. 110ff.; B. Croke, *Byzantion* 51 (1981) 126-31. Brightman believes that the approximate period of this legend (pontificate of Proclus, 434-46) probably marked the time for the insertion of the Trisagion into the liturgy. Indeed the earliest datable occurrence of the words comes a few years later during the Council of Chalcedon of 451 (*ACO* 2.1.1, p. 195, 30; cf. *ODCC*[3] 1642f., s.v. Trisagion; Quasten 58; Hanssens p. 110). It is possible, however, that the hymn originated earlier, as witnesses O. Zucker 36, an expanded Trisagion which L. Koenen has dated to the fourth century (Koenen 50f.; against this see Brock 30 n. 21; cf. also the historical continuum discussed by Koenen, p. 35). A Coptic legendary tradition in fact traces it to the apostolic period, claiming that it was revealed by the angels to Joseph of Arimathea and Nicodemus as they were burying Jesus (Ibn Siba, *Pretiosa margarita de scientiis ecclesiasticis* [ed. P.V. Mistih, Cairo 1966] p. 202f. [Arabic, with Latin trans. p. 499]; E.M. Ishaq, *The Coptic Encyclopedia* VII 2278 s.v. Trisagion; Brock 28; cf. further below p. 16 n. 46). For other accounts see Migne's preface to John of Damascus, *trisag.*, PG 95.17-20.

[23]The three cola reflect the principles of rhyme and isosyllabism:

| | **Syllables** | **Stresses** |
|---|---|---|
| ἅγιοϲ ὁ θεόϲ, | 6 | 2 |
| ἅγιοϲ ἰϲχυρόϲ, | 6 | 2 |
| ἅγιοϲ ἀθάνατοϲ | 7 | 2 |

The third, with its additional syllable, acts as clausula (cf. Engberding 171).

[24]The ἐλέηϲον ἡμᾶϲ seems to cast the hymn in *Du-Stil*. Many modern editors and translators (e.g., Brightman) have therefore interpreted the three ἅγιοϲ cola as vocatives: "Holy God, holy mighty, holy immortal," a reading which has the support of the Syrian and Georgian translations (Brightman 77; Hanssens p. 99; Quecke, *Stundengebet*, 301 n. 6). The difficulty with it is not the absence of vocative forms, which Engberding found so problematic, since ὁ θεόϲ in later (esp. Biblical) Greek is frequently a vocative (J. Wackernagel, *Über einige antike Anredeformen* [Programm Göttingen, 1912] 7-13; BDR § 147.2 n. 5; Mitsakis § 138 n. *; Gignac II 22 n. 6) but rather with the predicative position of ἅγιοϲ. Whenever ὁ θεόϲ and similar expressions function as vocatives, modifying adjectives normally stand in the attributive position (ὁ δεϲπότηϲ ὁ ἅγιοϲ, NT Rev.

E. Werner points out the frequency of epithet combinations such as "holy, " "strong," "eternal/immortal," in Jewish apocalyptic literature, such as the Apocalypse of Abraham.[25] He argues that, since the Byzantine doctors were well acquainted with the apocryphal and pseudepigraphic material, the Trisagion's synthesis of ἅγιοϲ, ἰϲχυρόϲ, and ἀθάνατοϲ must derive from this realm.[26] It is more likely,

---

6.10; ὁ θεὸϲ ὁ μέγαϲ καὶ θαυματὸϲ ἔπιδε ἐπὶ τοὺϲ δούλουϲ ϲου, *lit. Jac.*, Brightman 67.1; ἄκουε ὁ μέγαϲ θεὸϲ 'Αδωναῖε, *PGM* IV 1560; δεῦρό μοι, ὁ ἀκαταφρόνητοϲ θεόϲ, *PGM* VII 965f. Cf. LXX Jer. 39.18-19; NT Mk. 9.25, τὸ πνεῦμα τὸ ἄλαλον --- ἔξελθε. Further examples in Wackernagel op. cit.; A.N. Jannaris, *An Historical Greek Grammar* [London 1897] § 1251. *Suppl. Mag.* II 61.1 is, however, an exception: ἅγιοϲ ὁ θεόϲ, Καβριηλ, Μηχαηλ, ποίωϲαι τὴν ἡκανών μου [i.e., ποίηϲαι τὸ ἱκανόν μοι]). Thus, Engberding (168-74), Quecke (*Stundengebet*, 301 with n. 6), Koenen (*ZPE* 31 [1978] 76 with n. 6) and Brock (33 n. 17) argue that the syntax of the first colon requires "Holy is God," with the other cola following suit as "holy (and) mighty, holy (and) immortal"; i.e., a hymn in *Er-Stil*, like the song of the seraphim in LXX Isaiah 6.3. Evidence from elsewhere in the LXX supports this understanding (Is. 33.5, ἅγιοϲ ὁ θεόϲ, "Holy is God"; Ps. 98.9, ἅγιοϲ κύριοϲ ὁ θεὸϲ ἡμῶν, "Holy is the Lord our God") as does the syncretistic hymn known as the ἑνδεκάκιϲ ἅγιον (P. Berol. 9794), which alternates ἅγιοϲ ὁ θεόϲ with ἅγιοϲ εἶ (see the most recent ed. in R. Merkelbach, M. Totti, *Abrasax* 2 [Papyrologica Coloniensia XVII 2, Opladen 1991, pp. 123ff.]). Moreover, the Armenian version actually adds "and" between the two adjectives of the second and third cola (Brightman 423f.; Hanssens p. 99). As regards ἐλέηϲον ἡμᾶϲ, *Du-Stil* and *Er-Stil* patterns sometimes unite in the same hymn; cf. the ἑνδεκάκιϲ ἅγιον hymn cited above; Maas *FK* I 2.5-6: ἀρχαὶ καὶ ἐξουϲίαι, κυριότητεϲ λειτουργούϲιν *αὐτὸν* παριϲτάμεναι· Χερουβὶμ καὶ Cεραφὶμ τὰ πολυόμματα τὸν τριϲάγιον ὕμνον προϲφέρουϲι *ϲοί* (in general see Engberding 172). If this interpretation is correct, the liturgical version of the angelic Trisagion (see above, p. 7) probably reflects the same blending of styles: "Holy holy holy *is* the Lord of hosts, heaven and earth are full of *your* glory." Thus, structurally, the angelic and ecclesiastic hymns mirror each other.

[25]e.g., 27.8; 20.6; 22.1; 26.1 (cited by Werner 29 n. 1). This work is known only in the old Slavonic version, likely from a Hebrew original. See the Eng. trans. of R. Rubinkiewicz in J.H. Charlesworth (ed.) *The Old Testament Pseudepigrapha* I (Garden City, NY, 1983) 681ff.

[26]This apocalyptic milieu perhaps stands behind the joining of ἅγιοϲ and ἰϲχυρόϲ in descriptions of Jewish (and other) names of power in protective magical texts: Ιαω --- Ιαωουθ --- ἅγια κα⟨ὶ⟩ ἰϲχυρὰ καὶ δυνατὰ ὀνόματα

however, that the apocalyptic tradition as mediated through the New Testament Revelation[27] rather than through the Jewish writings accounts for the configuration of adjectives. The version of the angelic Trisagion in that book (4.8), sung to the enthroned deity by "the four creatures" (i.e., the cherubim[28]), closely corresponds to the epithets of the second and third cola of the ecclesiastic hymn: ἅγιοϲ ἅγιοϲ ἅγιοϲ | κύριοϲ ὁ θεὸϲ ὁ παντοκράτωρ[29] (i.e., ἰϲχυρόϲ), | ὁ ἦν[30] καὶ ὁ ὢν καὶ ὁ ἐρχόμενοϲ (i.e., ἀθάνατοϲ). In the verses which follow (9-11) the cherubim and the elders continue to worship τῷ ζῶντι εἰϲ τοὺϲ αἰῶναϲ τῶν αἰώνων (i.e., τῷ ἀθανάτῳ). (ὁ) ζῶν is elsewhere in Revelation an epithet of God (7.2, 10.6, 15.7) and Christ (1.18, see below), and among the honorific qualities which the angelic hosts ascribe to the enthroned Yahweh and the victorious Lamb is ἰϲχύϲ (5.12, 7.12).

Revelation also supplies the one use of ἰϲχυρόϲ to designate God in the NT; the LXX translators employ it frequently for this

---

τὰ τῆϲ μεγάληϲ Ἀνάγκηϲ, --- διαφυλάξατε ἀπὸ πάϲηϲ γοετίαϲ --- τὸ ϲῶμα καὶ τὴ⟨ν⟩ ψυχὴν --- Θωμᾶ, MacDaniel Phylactery 33ff. (IV/V AD; F. Heintz, *ZPE* 112 [1996] 296f.); similarly, ἅγια καὶ εἰϲχυρὰ καὶ δυνατὰ ὀνόματα, διαφυλάξατε Ἀλεξάνδραν ἀπὸ [π]αντὸϲ δεμονίου, Kotansky, *GMA* I 52.109ff. (IV AD); cf. Ιαω Cαβαωθ --- ἅγιον καὶ ἰϲχυρόν, κραταιὸν καὶ μεγαλοδύναμον {ον}ὄνομα, δὸϲ --- νίκην Πρόκλῳ, ibid. 58.10ff. (IV AD) and cf. parallels cited by Kotansky ad loc. Given the obscure origins of the Trisagion it is difficult to know whether these texts predate the hymn.

[27]For the relationship between the hymnic portions of Revelation and the later Jewish apocalyptic tradition, see R. Deichgräber, *Gotteshymnus und Christushymnus in der frühen Christenheit* (Göttingen 1967) 47-50, esp. 49f. for Rev. 4.8.

[28]Cf. Charles on 4.6 (I pp. 119-23).

[29]παντοκράτωρ is a translation for Hebrew Sabaoth, as it frequently is in prophetic literature; cf. Charles ad loc. (I p. 127); Brock 26; Bauer/Aland s.v.; further bib. in Daniel/Maltomini, *Suppl. Mag.* I 29.5-6 (n. p. 80). For ἰϲχυρόϲ as a rendering of a Hebrew divine name, see below n. 31.

[30]For the odd grammar, see below on lines 47-48, pp. 71f.

purpose.[31] ἀθάνατος however, for all its use of gods in Greek literature, never so occurs in the Greek Bible.[32] It is a significant point of comparison between archaic Greek and ancient Hebrew religious thought that the former articulated a parallel divine attribute negatively ("deathless"), the latter positively (חי, "living"; in Greek (ὁ) ζῶν, "(he) who lives"[33]). St. John of Damascus (c. 675 – c. 749) saw the correspondence, when in his *epistola de hymno trisagio* 3.39ff.[34] (Migne PG 95.29b) he derives the hymn's epithets from LXX Ps. 41.3,[35] τὸν θεὸν τὸν ἰσχυρὸν[36] τὸν ζῶντα, explaining ἀθά-

---

[31] ἰσχυρὸς κύριος ὁ θεὸς ὁ κρίνας, Rev. 18.8. ἰσχυρός was understood as a rendering of the Hebrew divine name אל and is so used 20 times in the LXX and regularly in Aquila (cf. Hier. *ep. ad Mar.* 25 [Migne PL 22.429]; Quell in *TDNT* III 79, 84 n. 88 s.v. θεός). Also in the LXX it describes God in his capacity as almighty judge (2Reg. 22.48, Ps. 7.12, Jer. 27.34) and appears frequently as an epithet in prayer (2Esdr. 11.5, 19.31, 32; Jer. 39.18; Dan. 9.4). All of these Septuagintal usages of the word well suit the atmosphere of the Trisagion, which celebrates the all-powerful *pantocrator*. It is used more generally of Christ, where the Baptist calls him ὁ ἰσχυρότερός μου (Mt. 3.11, Mk. 1.7, Lk. 3.16).

[32] ἀθάνατος is used of God in a *v.l.* to 1 Tim. 1.17 (cf. 6.16, ὁ μόνος ἔχων ἀθανασίαν) and never in LXX. The closest Biblical Greek ascription is ἄφθαρτος (Rom. 1.23, 1Tim. 1.17; twice only in LXX in Wisd. Sol.), also important as a divine epithet in Hellenistic philosophy and theology (see below n. 38; A.A. Long, D.N. Sedley, *The Hellenistic Philosophers* [Cambridge 1987] II 454; Harder in *TDNT* IX 96, 104 n. 53). It is a synonym (and probably of a related stem) to ἄφθιτος, which is mainly used of things in the archaic period, but later of gods (see M. Treu, *Glotta* 43 [1965/66] 7-10).

[33] For the frequency of this epithet in both the LXX and NT, see Bauer/Aland s.v. ζάω 1αε; Römer on *P. Köln* IV 172.6 (p. 50f.).

[34] B. Kotter, *Die Schriften des Johannes von Damaskos* IV (Berlin/New York 1981) pp. 289-332 (Migne PG 95.21-62).

[35] This interpretation is also given by Jobius Monachus (6th cent.; apud Phot. *biblioth.* 222, vol. 3 p. 181,5ff. Henry), who relates that the angels revealed the Trisagion directly to Proclus as a synthesis of Is. 6.3 and Ps. 41.3, and that the three epithets ἅγιος, ἰσχυρός, and ἀθάνατος distinguish the one true God from pagan idols; cf. Brock 28; Hanssens pp. 116f.

[36] τὸν ἰσχυρόν is read only in a few mss. Rahlfs omits it from the text of his Göttingen edition.

νατοϲ as a surrogate for ὁ ζῶν.[37] This interpretation may be more than the typical attempt to justify current ecclesiastical practice with Biblical proof-text.

Greek ἀθάνατοϲ and Hebrew/Semitic חי both set forth a relationship between the ideas of "deity" and "death." Whereas, however, the former describes gods as not suffering death, the latter means, among other things, the state of having conquered the powers of death, which may include the notion of having suffered death and risen, as in the case of Adonis, Esmun, Osiris, and others.[38] Such

---

[37]This identification could have been aided by the fact that in late popular usage ἀθάνατοϲ in certain contexts assumes a more positive force and approximates ζῶν or ζῷοϲ in meaning. For example, in papyrus nursing contracts ἀθάνατοϲ describes the state in which a child was given to a nurse and the condition in which she is expected to return it after her period of employment (C.Pap.Gr. I 4.26; 5.25; 7 ii.59; cf. the comments of Masciadri and Montevecchi in the intro. to this volume, pp. 27-29, and to their bibliography [28 n. 75] add Horsley, *New Docs* 2 p. 9). For similar usages in other kinds of papyrus documents, see O. Montevecchi, *La Papirologia* (Milan 1988) 219; J.H. Moulton, G. Milligan, *The Vocabulary of the Greek Testament* (London 1930) s.v. ἀθαναϲία. The Trisagion in the liturgy of the Abyssinian Jacobites apparently combined the two expressions: "Holy God, holy mighty, holy *living immortal*" (Brightman 218.1ff.; Hanssens p. 101).

[38]W.W. Graf Baudissin, *Adonis und Esmun* (Leipzig 1911) 450-510, esp. 492ff. and cf. 500 for the contrast between ἀθάνατοϲ and חי. Baudissin seems correct in his assertion that "living" as a divine epithet is non-Greek. *PGM* XII 79 ὁ ζῶν θεόϲ is a translation of Egyptian *ntr ʿnḫ* (see R.K. Ritner's n. ad loc. in *The Greek Magical Papyri in Translation* [Chicago 1992²], ed. H.D. Betz, p. 156 n. 19), a common designation in Egyptian texts, which the standard image of a god holding an *ankh* graphically represents (see A. Erman, H. Grapow, *Wörterbuch der aegyptischen Sprache* [Leipzig 1926-31; repr. Berlin 1971] I s.v. *ʿnḫ*, pp. 193-205, esp. on *ʿnḫj* [*ankhy*], "living one," a name of Re, Osiris, Horus, Thoth, etc. and cf. the feminine version for Hathor and Isis [p. 201]). Egyptian kings from Tutankhamon to the Ptolemies (after Philadelphos) bore the royal title and divine epithet "living image of the god Ammon" (e.g., *lapis Rosettanus* Grk 3 [*OGIS* 90], εἰκόνοϲ ζώϲηϲ τοῦ Διόϲ; cf. R.K. Ritner, *The Mechanics of Ancient Egyptian Magical Practice* [Studies in Ancient Oriental Civilization 54, 1993] 248 n. 1140). Jewish tradition makes rich and diverse use of the collocation "living God," especially in oaths and adjurations. Thus, various forms of adjuration in magical texts by "the living God" or "the name of the living God" probably

ideas underlie the application of ὁ ζῶν to Christ in, for example, Revelation 1.17b-18: μὴ φοβοῦ· ἐγώ εἰμι ὁ πρῶτος καὶ ὁ ἔϲχατος καὶ ὁ ζῶν, καὶ ἐγενόμην νεκρὸϲ καὶ ἰδοὺ ζῶν εἰμι εἰϲ τοὺϲ αἰῶναϲ τῶν αἰώνων καὶ ἔχω τὰϲ κλεῖϲ τοῦ θανάτου καὶ τοῦ ᾅδου. Here ὁ ζῶν signifies victory over death procured by Jesus through his resurrection.[39] Having defeated death, he possesses power over death. I think it likely that the Trisagion was originally dedicated to Christ with ἰϲχυρόϲ and ἀθάνατοϲ (i.e., ὁ ζῶν) as epithets of victory and apocalyptic kingship.[40] This explanation pro-

---

have Judaeo-Christian roots; e.g., Kotansky, *GMA* I 67.1ff., ὁρκίζω ὑμᾶϲ κατὰ τοῦ θεοῦ τοῦ {ζω} ζῶντοϲ (with n. ad loc.) and cf. NT Mat. 26.63, ἐξορκίζω ϲε κατὰ τοῦ θεοῦ τοῦ ζῶντοϲ. Cf. also *GMA* I 51.8f. with n. ad loc.; *PGM* IV 959, 1038, 1551ff.; A. Audollent, *Defixionum Tabellae* (Paris 1904) 247.18ff., 248A.1ff. On the other hand, *PGM* IV 559 ("Mithras Liturgy"), ϲύμβολον θεοῦ ζῶντοϲ ἀφθάρτου, may fall within the pale of Greek philosophic theology, and should perhaps be rendered, "symbol of the god who lives incorruptible," instead of with Betz, "symbol of the living, incorruptible god" (so Preisendanz' German). Cf. Antip.*Stoic*. 33 (*SVF* III 249,13f.), θεὸν τοίνυν νοοῦμεν ζῷον μακάριον καὶ ἄφθαρτον καὶ εὐποιητικὸν ἀνθρώπων, and esp. the Epicurean τὸν θεὸν ζῷον ἄφθαρτον καὶ μακάριον νομίζων (*ep*. 3 [*Men*.] 123 [Usener p. 59, 16f.]), where the grammar is clear (similarly, *phys*. 361, Usener p. 241, 26f.). In such combinations ζῶν/ζῷοϲ is not a true epithet, but makes predications of other epithets; similarly the Homeric θεοὶ ῥεῖα ζῶντεϲ (*Il*. 6.138; *Od*. 4.805, 5.122). It is also interesting to note that an etymology of the name of Zeus given by Cornutus is based on ζῆν (cf. K. Müller, *RE Supp*. IV 19 s.v. Allegorische Dichtererklärung).

[39]Frequently in the NT the resurrection is expressed simply through forms of ζῆν; e.g., Rom. 14.9: εἰϲ τοῦτο γὰρ Χριϲτὸϲ ἀπέθανεν καὶ ἔζηϲεν, ἵνα καὶ νεκρῶν καὶ ζώντων κυριεύϲῃ. Cf. Bauer/Aland s.v. [ζάω] 1αβ. The LXX renders Is. 26.19 "Thy dead shall live" with ἀναϲτήϲονται, while Aquila, Symmachus, and Theodotian all use ζήϲονται. See D. Hill, *Greek Words and Hebrew Meanings* (Cambridge 1967) 172. For other correspondences between the ideas of "life" and "victory" in Semitic thought, cf. R. Bultmann in *TDNT* II 840 n. 62 s.v. ζάω.

[40]Hanssens (pp. 112-116, esp. 115) approves the view of the Trisagion's origins expressed in the late Syriac *Book of Heraclides*. According to this work, which was attributed to Nestorius, the hymn was composed as a polemic against the Monophysites, since the Trisagion's epithets argue that only the human nature of Jesus suffered death, while his divine nature remained *impassibilis* (appar-

vides a clear rationale for the ecclesiastic Trisagion's coupling of these two adjectives in the context of its origins in Revelation's angelic Trisagion and other hymns of that book. It also brings the ecclesiastic hymn in close relation with the liturgical use of the angelic, which was frequently followed by the victory proclamation "Hosanna" (Mat. 21.9 and parallels) and called the ἐπινίκιος ὕμνος.[41] Another factor which supports this understanding is the pageantry surrounding the reading of the gospel in Oriental and Gallican liturgies, in which the ecclesiastic Trisagion was sung by a cleric during the procession of the gospel book, which represents Jesus' triumph over death.[42]

---

ently indicated by ἰσχυρός) and *immortalis* (ἀθάνατος). In response to this, Hanssens maintains, the Monophysites tried to claim the Trisagion for themselves by adding ὁ σταυρωθεὶς δι' ἡμᾶς to the ἀθάνατος colon (see below). I think it more likely that the hymn was later used by various factions in doctrinal controversy, but was not conceived for that purpose.

[41]*lit. Jac.* (Brightman 50.29); *lit. Mk.* (idem 131.29f.); *lit. Chys.* (idem 313.24); *lit. Pont.* (idem 522.12); *lit. Greg. Naz.* (Migne PG 36.708d); MPER N.S. XVII 53.2 with n. ad loc. (p. 96); cf. Brock 24; Werner 27. The question concerning the origins of this military epithet has caused some debate. Werner (loc. cit.) discounts the common theory which explains it on the grounds of the angelic Trisagion's association with the Hosannah, noting that the hymn is sometimes called ἐπινίκιος where the Hosannah is not added. In addition we may note that the NT Trisagion of Rev. 4.8 (cited above) has the cherubim singing the hymn τῷ καθημένῳ ἐπὶ τῷ θρόνῳ (4.9). This regal figure is the Father, but in effect the Son as well, since immediately before this scene Jesus speaks to the churches, encouraging them to be victors, ὡς κἀγὼ ἐνίκησα καὶ ἐκάθισα μετὰ τοῦ πατρός μου ἐν τῷ θρόνῳ αὐτοῦ (3.21). Jesus in fact is not distinguished from the enthroned Father until he appears as the Lamb in 6.16 and thereafter is closely associated with him. Thus the ἐπινίκιος epithet may have its roots in this earliest Christian angelic Trisagion, which celebrates the figure of *Christus Victor*.

[42]This is the interpretation of the rite given in the *expositio antiquae liturgiae Gallicanae* I 10-11 (ed. E.C. Ratcliff, London 1971, p. 7; see below p. 65f.). Cf. Quasten 60, 65f.; W.S. Porter, *The Gallican Rite* (London 1958) 27f.; A.A. King, *Liturgies of the Past* (London 1959) 160.

## The Trisagion—Trinitarian or Christological?

Since the Trisagion has its roots in the apocalyptic throne room where the victorious Lamb reigns coequal with the Father, it is not surprising that the hymn, in its original setting addressed primarily to Christ as victor and deity, also had strong Trinitarian affiliations. It is likely that the Trinitarian and Christological ascriptions were originally complementary and not viewed in opposition.[43] The storm of theological controversy over the nature of Christ eventually polarized the two perspectives and fostered the standardization of the Trinitarian interpretation in liturgies of the West and Orthodox East.[44] It is to this point we now turn.

The understanding of the Trisagion as an authoritative angelic proclamation concerning the nature of Christ encouraged various doctrinal factions to press it into their service.[45] Since the entire hymn refers to Jesus' glorified deity and kingship, Peter the Fuller (d. 488) and other Monophysites, through the addition of ὁ cταυρωθεὶc δι' ἡμᾶc[46] after the third colon of the hymn, promoted their belief that Christ had only one nature, the divine, and died as God, not man.[47] Reaction against the Monophysites may in fact have gen-

---

[43]This is indeed the case in our text; see below on the second of the three hymns, pp. 66-68. Cf. Brock 30f.

[44]Hanssens (p. 116) concludes, *Byzantinos, ut videtur, hymnum ad ss. Trinitatem ideo dirigere incepisse, ut omnem suspicionem nestorianismi a se arcerent et maxime, ut sibi 'monophysitas' patripassianismi et theopaschitismi accusandi facultatem compararent.*

[45]Koenen (34-44, esp. 34-38) has provided a comprehensive survey of these issues and I will recount them here only in so far as is needed for the understanding of our text. Cf. also Hanssens pp. 119-123.

[46]An important Egyptian legend of the Trisagion made these words part of the original hymn. According to it, angels appeared to Joseph and Nicodemus as they were burying Jesus and uttered the three cola of the hymn, to which they replied ὁ cταυρωθεὶc δι' ἡμᾶc. See above p. 9 n. 22.

[47]i.e., Theopaschism; cf. *ODCC*[3] 1305f., s.v. Theopaschites; W.H.C. Frend, *The Rise of the Monophysite Movement* (Cambridge 1972, repr. with corr. 1979) index p. 404 s.v. Theopaschism. This aspect of the controversy on Christ's nature is reflected in a fragmentary papyrus recently republished as

erated the exclusively Trinitarian application of the hymn. John of Damascus, one notable exponent of this perspective, extends the orthodox condemnation of Peter the Fuller's addition to an attack on the Christological application altogether: anyone who addresses the hymn to one of the *hypostaseis* instead of the Trinity κοινωνός ἐςτι τῆς τοῦ κναφέως τοῦ βαναύςου ςκαιότητος.[48] Thus he and other rigid Calcedonians championed the Trinitarian interpretation as the only safe and securely orthodox understanding.

While the efforts of theologians such as John of Damascus succeeded with regard to Byzantine and Western liturgies, they had no effect on theory and practice in Egypt, where both liturgies and popular compositions on papyri and ostraca preserved the Christological focus of the hymn. One reason for this is that Egypt was indeed a Monophysite stronghold. The mere fact of attribution to Christ, however, and the presence of ὁ ςταυρωθεὶς δι' ἡμᾶς does not necessarily establish a Monophysite ascription. Photios reports the argument of Ephraim, Patriarch of Antioch (527-45), that the use of this addition within the Christological Trisagion was a matter of regional preference, and the hymn in this form could be understood in an orthodox manner: φηςὶ δὲ ὁ Ἐφραΐμιος τὴν τοιαύτην ὑμνολογίαν τοὺς μὲν τὴν ἀνατολὴν οἰκοῦντας εἰς τὸν κύριον ἡμῶν Ἰηςοῦν Χριςτὸν ἀναφέρειν, καὶ διὰ τοῦτο μηδὲν ἐξαμαρτάνειν ἐπιςυνάπτοντας τὸ Ὁ ςταυρωθεὶς δι' ἡμᾶς, τοὺς δὲ Βυζάντιόν

*P.Carlini let.* 25, ed. A.C. Mancini (cf. esp. Horsley, *New Docs* 3, p. 111f., who also discusses related papyri). Cf. Anteon, *ep. ad Pet. Full.* 1.2: οὔτε γὰρ τὸ θεῖον λέγομεν παθητόν, ὡς cὺ φῄς, οὐδὲ γυμνὸν λόγον φαμὲν ςταυρωθέντα, ἀλλὰ Χριςτὸν Ἰηςοῦν, ἀλλ' οὐδὲ αὐτὸν μετὰ τῆς τριάδος. εἷς γὰρ τῆς τριάδος ἐςτὶν ἐνανθρωπήςας Ἰηςοῦς ὁ Χριςτός, ὁ δὲ ςταυρὸς περὶ τὸ ἀνθρώπινον τοῦ Χριςτοῦ γεγένηταί τε καὶ πιςτεύεται. ἐν δὲ τῷ τριςαγίῳ ποῦ ςώματος δήλωςις, ἵνα χώραν ἡ τοῦ ςταυροῦ λάβῃ προςαγορία; λέγεις γὰρ "ἅγιος ὁ θεός" (*ACO* 3 p. 217, 13ff.; cf. Koenen 35 n. 10). This quote establishes an important point on which orthodox and Monophysite agreed, that the Trisagion celebrates deity; it also raises the point of contention between the two camps, whether it is thus appropriate to introduce the crucifixion into a hymn which has this focus.

[48]*trisag.* 26.12f. (Migne PG 95.57a); cf. 5.23-36 (Migne PG 95.33a-b).

τε καὶ τὴν ἑσπερίαν νεμομένους εἰς τὴν ὑπερτάτην καὶ πανίερον πηγὴν τῆς ἀγαθότητος, τὴν ὁμοούσιον Τριάδα, τὴν δοξολογίαν ἀνάπτειν. δι' ὃ μηδ' ἀνέχεσθαι τούτους ἐπισυνάπτειν τὸ 'Ο σταυρωθεὶς δι' ἡμᾶς, ἵνα μὴ πάθος τῇ Τριάδι περιάψωσιν (*biblioth.* 228, IV p. 115 Henry). Ephraim's remarks are notable, not only for their advocacy of coexistence and toleration,[49] but also because they demonstrate that, although the question concerning Trinitarian versus Christological application became rooted in doctrinal disputes, on the level of popular usage geography, not controversy, exercised the greater influence.[50]

Further, when we examine Egyptian Trisagia, we find that often, in addition to ὁ σταυρωθεὶς δι' ἡμᾶς, other phrases were appended, e.g., ὁ σαρκωθεὶς δι' ἡμᾶς and ὁ ἀναστὰς ἐκ (τῶν) νεκρῶν. These attributes with the σταυρωθείς phrase give the Christological hymn a different complexion. Instead of exclusively referring to the Passion in a hymn which was strongly concerned with the deity of Christ, it became in effect a meditation on the entire person of Christ: his birth, death, and resurrection.[51] In this form, with these three (or similar) ascriptions, the Trisagion was in effect a hymnic expression of the second article of various Christian creeds.[52] We may take as a

---

[49]An attitude which was not too common. Photios stops short of censuring Ephraim for it, but says εἰκότως οἱ ἄλλοι πατέρες ἡμῶν παντελῶς περιεῖλον τὸ μὴ δεῖν προσάπτειν ἐν τῇ τρισαγίῳ δοξολογίᾳ τὸ 'Ο σταυρωθεὶς δι' ἡμᾶς, τῆς γὰρ τριαδικῆς θεολογίας ὕμνον εἶναι τὸ δοξολόγημα, ἐπεὶ καὶ ὁ Χερουβικὸς ὕμνος ἐν τῇ τριαδικῇ φωνῇ τῆς ἁγιότητος προερχόμενος, ἀπ' ἀρχῆς τῆς ὑπὲρ πᾶσαν ἁγιότητα καὶ ἀγαθότητα Τριάδος μελῴδημα ἐγνωρίζετο (ibid., IV p. 115f. Henry).

[50]See esp. Brock 29 with n. 18.

[51]See Koenen 37 and the examples cited on 38-44.

[52]Cf. Koenen 38; Lietzmann, *Symbolst.* 10ff./198ff. In an Ostracon edited first by F.E. Brightman (in W.E. Crum, *Coptic Ostraca from the Collections of the Egypt Exploration Fund*, London 1902, p. 4), then by Koenen (p. 42) a Christological Trisagion follows a Troparion, which is modeled after the Chalcedonian creed.

case in point the form of the hymn in the liturgy of the Coptic Jacobites[53]:

| | Syll. | Stress |
|---|---|---|
| ἅγιος ὁ θεός, ἅγιος ἰσχυρός, ἅγιος ἀθάνατος | 19 | 6 |
| ὁ ἐκ παρθένου γεννηθεὶς | 8 | 2 |
| ἐλέησον ἡμᾶς· | 6 | 2 |
| ἅγιος ὁ θεός, ἅγιος ἰσχυρός, ἅγιος ἀθάνατος | 19 | 6 |
| ὁ σταυρωθεὶς δι' ἡμᾶς | 7 | 2 |
| ἐλέησον ἡμᾶς· | 6 | 2 |
| ἅγιος ὁ θεός, ἅγιος ἰσχυρός, ἅγιος ἀθάνατος | 19 | 6 |
| ὁ ἀναστὰς ἐκ τῶν νεκρῶν | 8 | 2 |
| καὶ ἀνελθὼν εἰς τοὺς οὐρανοὺς | 9 | 2 |
| ἐλέησον ἡμᾶς· | 6 | 2 |
| | | |
| δόξα πατρὶ καὶ υἱῷ καὶ ἁγίῳ πνεύματι | 19 | 5 |
| καὶ νῦν καὶ ἀεὶ καὶ εἰς τοὺς αἰῶνας | 11 | 3 |
| ἁγία τριὰς ἐλέησον ἡμᾶς. | 11 | 4 |

In our hymn the "credal" formulae (all but the last positioned between the ἀθάνατος colon and ἐλέησον ἡμᾶς[54]) are ὁ ἀναστὰς ἐκ (τῶν) νεκρῶν (12, 38); ὁ σταυρωθεὶς δι' ἡμᾶς (25); ὁ ἐκ παρθένου σαρκωθεὶς θεός (52) and κατῆλθεν πρὸς ἡμᾶς καὶ σαρκωθεὶς ὑπὲρ ἡμῶν ἐκ τῆς ἀειπαρθένου καὶ βαπτισθεὶς ὑ-

---

[53]Brightman 155.11ff.; Hanssens p. 100. The liturgy of the Abyssinian Jacobites had an even more elaborate credal structure (Brightman 218.ff.; Hanssens p. 101f.; Quasten 59). The Christological focus is also indicated by the position of the Trisagion in Egyptian liturgies, i.e., before the gospel instead of before the lections in general (as in the Byzantine, Nestorian, and Armenian rites); cf. A. Chupungco in *EEC* II 853 s.v. Trisagion. The Gallican tradition employs it at both places (Quasten pp. 60, 65f.; cf. above p. 15 with n 42). For the rhythm, cf. the "synagogue prayer" cited above p. 4f.

[54]Thus, the grammatical configuration of these ascriptions (except the last) is that of a protasis in the form of a participle with a definite article (which functions as a vocative: "you who") and an imperatival apodosis. On this style see A. Dihle, *BZ* 69 (1976) 2.

πὲρ ἡμῶν (68-70).[55] Of course, the presence of such phrases with ὁ σταυρωθείς does not preclude a Monophysite label. Other language in our hymn may actually be considered as such (e.g., Ἡρῴδης ἐσπούδασεν θεὸν ἀποκτείνειν [10f.]; Ἰωάννης ἐβάπτισεν κύριον τὸν θεὸν ἡμῶν [14f.]; ọ γὰρ πάντων δεσπότης πεῖραν ἔλαβεν [23f.]), but these ascriptions may also be understood in the context of the high Christology of Athanasian orthodoxy.[56] The hymn's provenance of Egypt makes likely that we have before us a Monophysite document, but nothing in its content necessitates it.[57] In addition, the second hymn of our text integrates Trinitarian conceptions into a chiefly Christological framework,[58] as does the last three lines of the Jacobite hymn cited above.

[55]Of course, orthodox theologians such as John of Damascus censured all such predications: τότε εἰς τὸν υἱὸν μόνον φαμὲν ὕμνον λέγεσθαι, ὅτε τὰ ἐνδεικτικὰ τῆς ὑποστάσεως αὐτοῦ λέγομεν, οἷον, ὁ υἱὸς τοῦ θεοῦ, ὁ λόγος, ἡ σοφία τοῦ θεοῦ καὶ δύναμις, ὁ ἐκ πατρὸς γεννηθείς, ὁ σαρκωθείς, ὁ σταυρωθεὶς σαρκί, ὁ ἀναστάς, ὁ ἀνελθὼν εἰς οὐρανούς, ὁ καθήμενος ἐν δεξιᾷ τοῦ πατρός, καὶ τὰ τοιαῦτα. ὥσπερ οὖν ταῦτα ἐνδεικτικὰ τῆς μιᾶς ὑποστάσεως καὶ οὐδαμῶς ἐφ' ἑτέρας ὑποστάσεως ἢ ἐπὶ τοῦ κοινοῦ τῆς θεότητος χώραν ἔχει, οὕτω καὶ ὁ τρισάγιος ὕμνος ἐνδεικτικὸς ὢν τῶν τριῶν ὑποστάσεων ἐπὶ μιᾶς τῶν ὑποστάσεων χώραν οὐκ ἔχει (*trisag.* 3.6-14 [Migne PG 95.28b-c]).

[56]Fr. Justin's remarks are apt at this point: "One can make so bold as to say, 'God lay in a manger, God wearied, hungered, thirsted, suffered, was crucified and died,' and still be understood in an Orthodox manner. At the same time, we know that the *Godhead* cannot suffer, be crucified, or die." In his examination of this document neither he nor Fr. Haralampos were willing to label it Monophysite (Correspondence dated 7 September 1991; see preface). Cf. also Hanssens p. 115, *Re tamen vera neque necessario monophysismo adhaeret, qui affirmat Deum passum et mortuum esse, neque nestorianismo qui confitetur Deum esse impassibilem et immortalem. Utrumque enim plane orthodoxe asseveratur.*

[57]The fact that "orthodox" and "Monophysite" camps contain within themselves numerous theological gradations and shadings further complicates the issue. With regard to the latter, for example, M. Simonetti observes that after the second half of the 5th cent. we must distinguish between the more extreme "real Monophysism" and "verbal Monophysism" (*EEC* I 569b s.v. Monophysism). Cf. also E. Klum-Böhmer, *Das Trishagion als Versöhnungsformel der Christenheit* (München/Wien 1979) chh. 3, 4 and p. 70.

[58]See introduction to this section, below p. 67f.

## *P. Mich.* 799: Theological and Liturgical Context

Although the three hymns of our document, taken as a whole, seem at first glance loosely strung together, they reveal on closer inspection a unified theological perspective which precedes along well-defined and familiar grooves. The first hymn expresses worship to Jesus Christ through a recounting of his birth, baptism, suffering, and victory over sin and death. The second focuses on the heavenly throne room to which he has been restored[59] via Isaiah's vision of the seraphic praise of the Trinity and the worship of Jesus as incarnate savior and creator. The third continues in that vein, celebrating Christ as cosmic creator through appropriate Septuagintal allusions. The concluding sextuplet enjoins the faithful to contemplate his φιλανθρωπία, returning in ring composition fashion to the incarnation and baptism (κατῆλθεν πρὸϲ ἡμᾶϲ, καὶ ϲαρκωθεὶϲ ὑπὲρ ἡμῶν ἐκ τῆϲ ἀειπαρθένου καὶ βαπτιϲθεὶϲ ὑπὲρ ἡμῶν, 68-70).[60] Thus the total document unites the work of Christ as creator, *pantocrator*, and incarnate savior, as often in orthodox logos theology: the eternal logos, who was the principle by which the kosmos was created, must also become flesh, i.e., participate in the created order, and redeem it.[61] On another level the cycle of hymns, by juxtaposing the humility of the incarnate Jesus and the glory of the *pantocrator*, reflects the *kenosis*/exaltation motifs as classically stated in NT Phil.

---

[59]Cf. Jn. 17.5, where Jesus prays, now that his earthly mission is accomplished, καὶ νῦν δόξαϲόν με ϲύ, πάτερ, παρὰ ϲεαυτῷ τῇ δόξῃ ᾗ εἶχον πρὸ τοῦ τὸν κόϲμον εἶναι παρὰ ϲοί. The δόξα of which he speaks is that of the eternal logos-creator (cf. Jn. 1.1-18).

[60]The first hymn also has a kind of "ring composition" structure. See below pp. 45f., 66.

[61]See, e.g., Ath. *de inc.* 1.4 Kannengiesser, SC 199 (Migne PG 25.97c): οὐδὲν γὰρ ἐναντίον φανήϲεται, εἰ δι' οὗ ταύτην (sc. τὴν κτίϲιν) ἐδημιούργηϲεν ὁ Πατήρ, ἐν αὐτῷ καὶ τὴν ταύτηϲ ϲωτηρίαν εἰργάϲατο. Cf. also Koenen 44 on lines 16-26 of the aforementioned ostracon (above n. 52).

2.6-11[62]. Because Jesus emptied himself, taking the form of a servant and becoming obedient unto death, God exalted and glorified him, giving him the supreme name of power. Indeed the line of dashes across the top margin of the hymn beneath the *nomen sacrum* IC XC may serve to establish at the outset what soon becomes obvious, viz., that all three hymns, not just the first, concern Christ, by presenting a united doctrinal focus consisting of logos and kenotic motifs.

It is difficult to determine the original nature and purpose of *P. Mich.* 799. To begin by way of exclusion, we cannot class it among the private prayers, such as *P. Oxy.* III 407, VII 1058, 1059, which are usually in the first person and at times include the name of the petitioner.[63] At the other end of the spectrum, it is not a formal liturgy sanctioned for general public use. Save for credal language and the Trisagion stanzas themselves, it cannot be reproduced in such liturgical traditions known to us. On the other hand, the injunctions of the concluding sextuplet, δεῦτε οὖν πάντεϲ οἱ πιϲτοί, ἴδετε καὶ θαυμάϲατε καὶ ⟨ ⟩ τὴν ἀνεκδιήγητο⟨ν⟩ ἑαυτοῦ φιλανθρωπία⟨ν⟩ (65ff.), suggest communal worship or liturgy of some kind.[64] I here set forth three possible interpretations, the first of which, in my opinion, has the most to commend it.

Baptismal language and imagery loom large in the entire piece, including the mentioning of Jesus' baptism by John early in the narrative of the first hymn (14f.) and the designation of this event as part of his *katabasis* in the final sextuplet (70). In addition to these two

[62]ὃϲ ἐν μορφῇ θεοῦ ὑπάρχων --- ἑαυτὸν ἐκένωϲεν --- ἐταπείνωϲεν ἑαυτὸν --- διὸ καὶ ὁ θεὸϲ ὑπερύψωϲεν καὶ ἐχαρίϲατο ἀυτῷ τὸ ὄνομα τὸ ὑπὲρ πᾶν ὄνομα. For a thorough discussion, see Martin op. cit. (above p. 6 n. 14) 165-96; cf. A. Oepke, *TDNT* III 661 s.v. κενόω.

[63]Our hymn does, however, have in common with texts of this kind a strong penchant for the adaptation of LXX material to hymns. Cf. in general Schermann 196ff. and above p. 1 n. 3. For Trisagia in the private prayers of monks, see Koenen 39 with n. 26.

[64]On this see also the preface to the second hymn, below p. 66.

specific references, which, coupled with the incarnation theme, structurally frame the entire document (see above p. 21), our text applies to Jesus Old Testament portraits of Yahweh subduing and restraining the waters of chaos (ὁ περιπατῶν ὡϲ ἐπὶ ξηρᾶϲ ἐπὶ θαλ{λ}άϲ⟨ϲ⟩ηϲ, 57 [Job 9.8]; ὁ τιθεὶϲ ⟨ἄμ⟩μον ὅριον ἐν τῇ θαλ{λ}άϲ⟨ϲ⟩ῃ{ϲ}, 64 [Jer. 5.22]). Such ascriptions admit of two levels of Christological interpretation: Jesus as creator/*pantocrator* and typologically as the one who purges and sanctifies the baptismal waters (see ad locc.). Third, apart from specific reference to and scriptural typology of Jesus' baptism, our document (especially the second hymn) contains a number of Biblical references and other material which are employed or which are similar to those employed in the Egyptian/Coptic and Greek liturgies of baptism.[65] These include (see commentary for discussion): μαρτύρων καὶ προφητῶν (18); εἰϲ μίαν οὐϲίαν ⟨καὶ⟩ κυριότητα (45); ὁ κλίναϲ τοὺϲ οὐρανοὺϲ καὶ κατῆλθεν πρὸϲ ἡμᾶϲ (47-48); ὁ πλάϲαϲ τὸν ἄνθρωπον κατ' ἰδίαν εἰκόνα (50-51); ὁ ἐκτείναϲ τὸν οὐρανόν (56); τὴν ἀνεκδιήγητο⟨ν⟩ ἑαυτοῦ φιλανθρωπία⟨ν⟩ (67).

The combination of these three features rather than any one of them suggests the connection of *P. Mich.* 799 to some baptismal context. Specifically, our document may be a cycle hymns for the Epiphany, which in the East continued primarily to commemorate the baptism of Jesus and secondarily his incarnation.[66] It could also be a

[65]The authoritative treatment of the Coptic liturgy of Baptism is that of Burmester, "Baptismal Rite" 27-86. The same author's *Coptic Church* 111-126, although abbreviated from the former work, gives fuller information on certain particulars. More recently, cf. Abp. Basilios in *The Coptic Encyclopedia* I 339-343 s.vv. "Baptism, Liturgy of." The two works of Burmester translate much of the Coptic text, but for a fairly complete translation, cf. Woolley, *Coptic Offices* 1-58. For the Greek baptismal rite, see Goar, *Euchologion* 287-306.

[66]In the West it eventually focused on the visitation of the Magi. On the celebration of the Epiphany in general (6 Jan. ) cf. *ODCC*[3] 554; V. Saxer in *EEC* I 282; K. Treu has identified as Epiphany hymns P. Berol. inv. 1163 (V/VI; *APF* 21 [1971] 62-65) and P. Berol. inv. 16595, (V/VI; *APF* 24/25 [1976] 121-23). On these texts see also Horsley, *New Docs* 2, p. 164f. Examples from the Byzantine tradition, cited by Treu, include W. Christ, M. Paranikas, *Anthologia Graeca*

composition to accompany the liturgy of baptism.[67] The association of a Trisagion hymn with the baptismal office meshes with what we know about Egyptian practice. Both types of Trisagion play a role in the Coptic baptismal rite: the ecclesiastic hymn is recited before the "Prayer of the Gospel" during the preliminary service of the blessing of the basin and the post-baptismal ceremony of the loosing of the girdle[68]; the angelic version is sung during the exorcism and blessing of the waters of the font.[69]

At the same time, I must here set forth two caveats. First, many of the echoes of the baptismal liturgy in this hymn are not confined to that context, but surface in other quarters of the worship life of east-

---

*Carmina Christianorum* (Leipzig 1871; repr. Hildesheim 1963) Kathismata II γ′, IV β′, VI γ′ (pp. 57, 58, 61).

[67]If this correct, the occasion may have been the general baptism, which since the 4th/5th centuries was celebrated in the Egyptian church on Saturday of the fourth week of Lent. In addition to this local custom, the Egyptians may also have recognized the more universal general baptism on Easter eve. See Burmester, "Baptismal Rite" 82f.

[68]Burmester, "Baptismal Rite" 49; idem, *Coptic Church* 125; Woolley, *Coptic Offices* 50.

[69]Burmester, "Baptismal Rite" 56, 73; idem, *Coptic Church* 121; Woolley, *Coptic Offices* 36. The angelic Trisagion enjoyed considerable esteem as a formula of exorcism (see above p. 8 n. 21). Thus, it appropriately occurs in a specifically exorcistic section of the baptismal liturgy, in which the waters of the font are purged and blessed. Cf. earlier in this same section of the Coptic rite the following prayer, which is not paralleled in the Greek ceremony: "Creator of the waters, Maker of the universe, we call upon thy holy and eternal power, the name which is above every name, of thine only-begotten Son Jesus Christ our Lord, who was crucified for us under Pontius Pilate. We pray thee, our Master, for thy servants; change, transform, hallow them, give them power; that through this water and this oil, all adverse powers may be brought to nought; all evil spirits restrain, confound, and render helpless; all magic, all sorcery, all idolatry, all witchcraft bring to nought" (trans. Woolley, *Coptic Offices* 33). For other exorcistic elements, cf. Burmeister, "Baptismal Rite" 62, 64, 68f., 72, 74. More generally on exorcism and baptism, see K. Thraede in *RAC* VII 76-100 s.v. Exorzismus; R. Merkelbach, *Abrasax* 4 (Papyrologica Coloniensia XVII 4; Opladen 1996) pp. 9-13, 43.

ern Christianity[70] (see on the *Apodeipnon* below). Second, baptismal language and imagery do not limit a hymn to a baptismal *Sitz im Leben*. Baptism in late-antique Christianity provided sufficiently fertile ground for both hymnody and theological reflection without demanding a specific association with that liturgical realm. For example, although Maas *FK* I 5 evinces a considerable presence of baptismal themes,[71] it is not a hymn connected with either baptismal context above mentioned, but rather with the *Apodeipnon* tradition.

Our document in fact has much in common with hymns of the ἀπόδειπνον,[72] the evening office of eastern churches which corresponds to the western Compline. Common elements (which, however, are by no means unique to the ἀπόδειπνον) include focus on the angelic adoration of the Trinity as the supreme model of worship (our text, 40ff.; cf. *FK* I 1.1-5, 2.5-6, 3.1-8); martyrs and prophets as important players in the liturgical drama (17f.; cf. *FK* I 1.7, 2.7-9, 3.9-10); invitation to worship by δεῦτε (πάντεϲ πιϲτοί) (65; cf. *FK* I 2.1, 3.1); and emphasis on φιλανθρωπία as a divine attribute which incites human devotion (67; cf. *FK* I 3.11, 20, 6.3; Trypanis, "Three Hymns," # 1 passim).[73] Our hymn does not occur in any of the extant *horologia* as part of the ἀπόδειπνον, but this is of little consequence, since hymns used in the evening office were not fixed,

---

[70]For example, the above cited phrase in our hymn "who bowed the heavens and came down" (47-48) does indeed occur three times in the Coptic baptismal office (see below ad loc.) but as the second of the "the three absolutions" it also appears in numerous liturgies. See Burmester, *Coptic Church* 330ff., and his index s.vv. Three Absolutions, Prayers of the" (p. 388).

[71]This hymn, like our first hymn (1-39), presents an account of the earthly life and glorification of Christ. The narrative of his birth in *FK* I 5.9-11 especially bears resemblances to that of our text (see commentary to lines 1-11 below). The hymn's baptismal themes (12-17) are similar to those of the Epiphany hymn of P. Berol. inv. 1163; see Treu *APF* 21 (1971) 63.

[72]See the Byzantine examples in Maas, *FK* I 1-6 and Trypanis, "Three Hymns"; for further discussion and bibliography on this material see A. Dihle, *BZ* 69 (1976) 1ff. and D. Hagedorn, *ZPE* 52 (1983) 275 n. 1.

[73]For these and other points of similarity, cf. comm. below on lines 1-2, 10-11, 14-15, 17-18, 65, 66-67.

but varied from location to location.[74] Thus *P. Mich.* 799, especially the first hymn, may be an evening song with a Trisagion structure imposed upon it.[75] A synthesis of this kind would not be out of place, since the Egyptian evening office regularly included the Trisagion.[76]

We also know, however, that theological hymns were employed as didactic tools on the popular level, both to advance heresy and to oppose it. For example, the heretic Bardaisan (AD 154-222)[77] promoted his teachings among the people through 150 hymns (intended to correspond to the Hebrew Psalter)[78] composed in a style, proba-

---

[74]For the *horologia* of the Egyptian/Coptic tradition, see Quecke, *Stundengebet* 13-52. For the Egyptian ἀπόδειπνον in particular and for its wide variance in content in the different *horologia* mss., see ibid. 32-34. Similarly with regard to Byzantine ἀπόδειπνα, Maas notes that of the 17 *horologia* mss. and editions which categorize *FK* I 1-6 as ἀπόδειπνον hymns, one has all six, another five, three others three, all of which have ἡ ἀcώματοc φύcιc (*FK* I 1) as the first in the series (Maas, *FK* p. 3; idem, *BZ* 18 [1909] 310f., 319). The 12 others have only one hymn: ten have ἡ ἀcώματοc φύcιc, two have another hymn which is # 6 in Maas' collection. This evidence suffices to prove the flexibility of the tradition.

[75]Perhaps the most striking stylistic difference between the ἀπόδειπνον hymns and our text is the numerous references of the community to itself in the first-person plural in the former: δεῦτε πάντεc πιcτοὶ προcκυνήcωμεν, *FK* I 2.1; esp. in requests that the righteous dead pray for them: ἅγιοι μάρτυρεc, δυcωπήcατε τὸν μόνον φιλάνθρωπον --- ἱλαcμὸν δωρήcαcθαι ταῖc ψυχαῖc ἡμῶν, *FK* I 3.10ff.; and so frequently. In some evening hymns monks make reference to their previous or ensuing sleep (e.g., Trypanis "Three Hymns," 1.5f., 2.11ff.).

[76]Quecke, *Stundengebet* 32. On the issue of *contaminatio* see below p. 44f.

[77]In general see H.J.W. Drijvers, *Bardaisan of Edessa* (Studia Semitica Neerlandica 6 [Assen 1966]); idem (ed.), *The Book of the Laws of Countries: The Dialogue on Fate of Bardaisan of Edessa* (Assen 1965). His thought may be accurately described as a form of Gnosis; see K. Rudolph, *Die Gnosis* (Göttingen 1990³) 353f. with nn. 154 and 155 for further bibliography (Eng. trans. [San Francisco 1987] 327ff. nn. 154 and 155).

[78]Ephraim Syrus mentions only Bardaisan as the author of these hymns; cf. the fragments collected by Ephraim in his *hymn. contra haer.* 55 (ed. E. Beck, *Des Heligen Ephraem des Syrers Hymnen contra haereses* [Louvain 1957, Corpus

bly invented by him, called the *madrasa*.[79] To refute his heresy, Ephraim Syrus (c. 306-73) composed orthodox hymns in the same form, of which he became master, and directed choirs of nuns to sing them outside of churches.[80] It is possible that our hymn, as well as other Trisagia and hymns on papyri and ostraca, be placed within this tradition of popular didactic poetry. It is easy to see how the Trisagion, having long been a medium for the expression of theological ideas, on the popular level could have become a catechizing tool for various theological groups.[81] In conclusion, however, I will say that this didactic model and the baptismal interpretation fit nicely together. Although Egyptian Christians commonly practiced infant baptism

---

Scriptorum Christianorum Orientalium vol. 169 pp. 207ff., Syriac; vol. 170 pp. 186ff. trans.]; Eng. trans. in M. Sprengling, *The American Journal of Semitic Languages and Literatures* 32 [1916] 196-98). According to Sozomenos (*h.e.* 3.16.5-7) his son Harmonios wrote them (cf. *Selections from the Hymns and Homilies of Ephraim the Syrian*, trans. by J. Gwynn [The Nicene and Post-Nicene Fathers XIII, 2nd ser., 1905] pp. 129f.; cf. also Theodoret, *h.e.* 4.29.2). This contention, and even Harmonios' existence, has come into doubt, in part due to the over-appropriateness of his name (K.E. McVey, *Ephrem the Syrian – Hymns* [New York 1989] 26 n. 103; Sprengling, op. cit., 199-202). Harmonios probably did exist; the name may actually be a translation of a Semitic name (Drijvers, *Bardaisan* [cf. above n. 77] 143 n. 2) and Ephraim himself in another work recognizes and quotes a son of Bardaisan without, however, naming him (C.W. Mitchell, ed., *S. Ephraim's Prose Refutations* II [Oxford 1921] p. cv [transl.], p. 222f. [Syriac]). On the question of the authorship of the hymns, Ephraim's testimony is probably to be trusted over Sozomenos' (Drijvers, *Bardaisan* 149f.).

[79]The structure of the *madrasa* is indeed based on many of the same principles of later Greek Byzantine church poetry, i.e., isosyllabic verse, parallelism, rhyme, alliteration, and various kinds of word play (cf. McVey op. cit. [above n. 78] 26). The similarities, however, do not necessitate the conclusion of certain modern scholars that the former begot the latter or vice versa, as Sozomenos claimed (loc. cit., above n. 78). Both forms are explainable from poetic principles in their respective languages (cf. above, p. 3 n. 11).

[80]Sozomenos loc. cit. (above n. 78); I owe this observation to Fr. Justin (see Preface above).

[81]The instructional value of hymns was recognized from a very early stage; cf. NT Col. 3.16: διδάϲκοντεϲ καὶ νουθετοῦντεϲ ἑαυτούϲ, ψαλμοῖϲ ὕμνοιϲ ᾠδαῖϲ πνευματικαῖϲ.

since the fifth century, the conservatism of the Coptic liturgy preserved to a higher degree than others the older catechetical nature of the baptismal office, in which the neophyte was instructed in the faith before he entered the waters of the font.[82]

---

[82]See Burmester, "Baptismal Rite" 42f. and cf. 61; idem, *Coptic Church* 111 (and in general cf. his Preface p. xi).

## II. TEXT OF *P. MICH.* 799

Inv. no. 4936 23.3 x 38.5 cm 7th cent.

Provenance unknown

The leather, orange-brown in color, with a crest at the top and a smaller one at the bottom, slightly tapers from the widest point at the top (23.3 cm) to bottom (19.7), and is unevenly cut on both sides. This "keystone" shape may have facilitated rolling up the document as a scroll.[83] The surface has sustained damage from water or some other liquid on the upper part, especially on the right side, resulting in some letters at the ends of lines 8 and 9 being partially washed away. The margins measure on the left .4-1.7 cm; right 0-4.2 cm; bottom .5-1.2 cm; top (measured to the top bar of IC XC) 1.3-4.6 cm. At the top and slightly left of center the *nomen sacrum* IC XC is separated from the rest of the text by a line of dashes running the full width. The hand resembles other literary and ecclesiastical scripts of the sixth and seventh centuries, such as W. Schubart, *Papyri Graecae Berolinenses* (Bonn 1911) 44b, 48b; and especially G. Cavallo, H. Maehler, *Greek Bookhands of the Early Byzantine Period* (London 1987) 53a, 53b, 53c.[84] The flesh side is blank.

The first and second hymns (see discussion of content in intro. above) begin and end with crosses (dipl. lines 2, 14, 15, and 22).[85]

---

[83]"Although the shape of the manuscript when opened seems to have no significance, if we think of it as a scroll, it does become very significant. The text would have been rolled up from the bottom. The slight taper to the sides would ensure that the inner parts of the manuscript were tucked neatly under the outer, and thus protected. The point at the top would form a tab, a convenient place to tie the scroll and thus keep it together" (Fr. Justin [see preface], correspondence dated 10/15/94).

[84]L.S.B. MacCoull discusses the hands of these three documents in her publication of another text of similar hand and date, P. Copt. Mus. inv. 3469 (*APF* 40 [1994] 131f.). I thank her for a preliminary copy of this article.

[85]With this well-known Byzantine/Christian convention (on which see M. Naldini, *Il Cristianesimo in Egitto* [Fiesole 1998$^2$] p. 27) cf. in secular docu-

Those that begin the two hymns are the simple "plus sign" type; those that close each hymn are a more elaborate, cursive form.[86] In the first hymn the scribe punctuates most cola of the Trisagion proper with //[87] but abandons this practice in the second and third hymns and dispenses with the crosses in the third, as space becomes tight and the writing more cramped. A blank space of about eight letters follows the second cross at the end of dipl. line 14, marking the end of the first hymn.

Our scribe makes sparing use of abbreviated *nomina sacra*, employing them only with the titular IC XC[88] and with ΘC in all of the first cola of the Trisagion refrain save one.[89] In all other cases he writes out divine names which are frequently abbreviated in Christian documents.[90] Other paleographic features include καί written as ϛ

ments the occasional use of small x-like markings to close a text (*P. Mich.* III 194; *P. Lund* VI 2 [*SB* VI 9352]; *P. Oxy*. LXI 4115.4 with n.; H.C. Youtie, *Eos* 48, 1956, 390 [*Scriptiunculae* II 906]).

[86]In the diplomatic I distinguish these two forms with + and † respectively. For the actual appearance of the cursive form, see the plate and the images accompanying the diplomatic text which follows, and cf. MPER N.S. XVII 12 end (pl. 8); 38.1 (pl. 25); 54 beginning of each line (pl. 37).

[87]Diplomatic lines 2, 3, 4, 6, 7, 8 (?), 10, 13 (in two of the three places where these strokes do not appear [lines 11 and 12] there are small spaces after the cola). Similar strokes are used for marking verses in poetical texts which are written continuously, such as *P. Lond. Lit.* 52 (III AD), 53 (III AD) and the school exercises *P. Ross. Georg.* I 12 (II/III AD) and 13 (III AD). Cf. also strokes or dashes to mark sense breaks: *P. Oxy.* L 3533 (III AD, literary); MND 552 E-F (B. Boyaval, *ZPE* 17 [1975] 145-50 [IV AD, Biblical]). Such markings in these texts and ours may have served to assist reading and/or recitation; cf. Turner, *GMAW* p. 144 on text 86 (*P. Oxy.* 3533); idem, *The Typology of the Early Codex* (Philadelphia, 1977) 84-85 with n. 14; and especially R. Cribiore, *GRBS* 33 (1992) 259-63, who also discusses the use of crosses for similar purposes.

[88]This name has strokes both above and below the letters, a configuration for which I find no parallels.

[89]3, 16, 29, 55 (dipl. 2, 6, 11, 22); the exception is 46 (dipl. 18).

[90]Forms of θεόc (5, 11, 15, 46, 53; dipl. 3, 5, 6, 18, 22); Ἰηcοῦ⟨c⟩ (4; dipl. 3); κύριον (15; dipl. 6). The Michigan Trisagion also does not abbreviate οὐρανόc (1, 33, 47, 56; dipl. 2, 12, 18f., 23) or ἄνθρωποc (50 dipl. 20) which

(dipl. 13, 25 [bis], 26, 27, 29 [bis], 30, 31)[91] and trema in ελη̈ϊcων (14 dipl.) and ϊδετε (29 dipl.).[92]

(The digitized images which face the following diplomatic transcript were scanned at 600 dpi and considerably reduced. For a photo of the complete text, see plate.)

---

some Christian papyri treat as *nomina sacra* (L. Traube, *Nomina Sacra* [München 1907] 99-103; A.H.R.E. Paap, *Nomina Sacra in the Greek Papyri of the First Five Centuries AD* [Papyrologica Lugduno-batava 8, 1959]104-06). For other Biblical/Christian texts which use abbreviated divine names inconsistently, cf. Horsley, *New Docs* 3, 112 (with refs. there cited), 116; Paap. op. cit. 100-13 passim, esp. 101 n. 4. The latter author's contention, however, that scribes who usually write out divine names do so from ignorance of abbreviations, cannot hold for our text, where θεόc and 'Ιηcοῦc are both abbreviated and written fully.

[91] I take ϟ as a very cursive and rapid writing of καί rather than an actual abbreviation. See H.C. Youtie, *CP* 33 (1944) 34 n. 64 (*Scriptiunculae* II 836 n. 64).

[92] Cf. Turner, *GMAW* p. 10f.

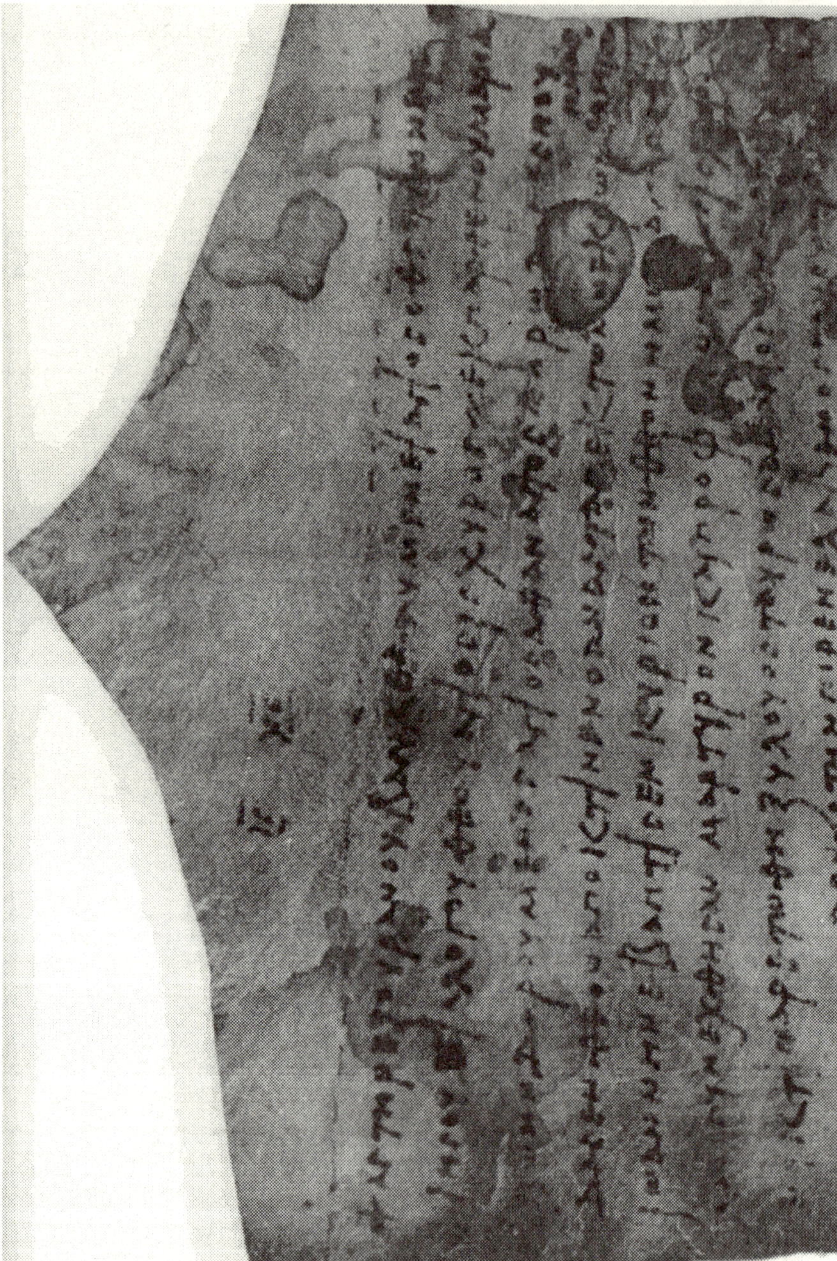

## Diplomatic Text

I̅C̅ X̅C̅

— — — — — — — — — — — — — — — — — — — —

+αcτηρεξουρανουβαcιλεαcυμενειαγιοcοθ̅c̅//γεννατε
ιηcουδιαλογουθεουαγιοcιcχυροc//εκπαρθενουμαρια
ζωηνδωρουμενοcαγιοcαθανατοc//ηρωδηcεcπου
δαcενθεοναποκτινενοαναcταcεκτωννεκρωνεληιcω`ημαc´
ιωαννηνεβαπτιcενκυριοντωνθεονημωναγιοcοθ̅c̅//
λαοιcυνεχθηcαν μαρτυρονκαιπροφητωναγιοcιcχυροc//̣
νυκτιπαρετωθηξυλουοcταυρωcαναγιοcαθανατοc//̣ο̣

---

2 line over *nomen sacrum* extends over // with what seems to be an insertion point above the ω  5 ημαc written above εληιcω  6 line over *nomen sacrum* begins over ο and extends to edge of parchment  8 ink at end of line disfigured through water damage

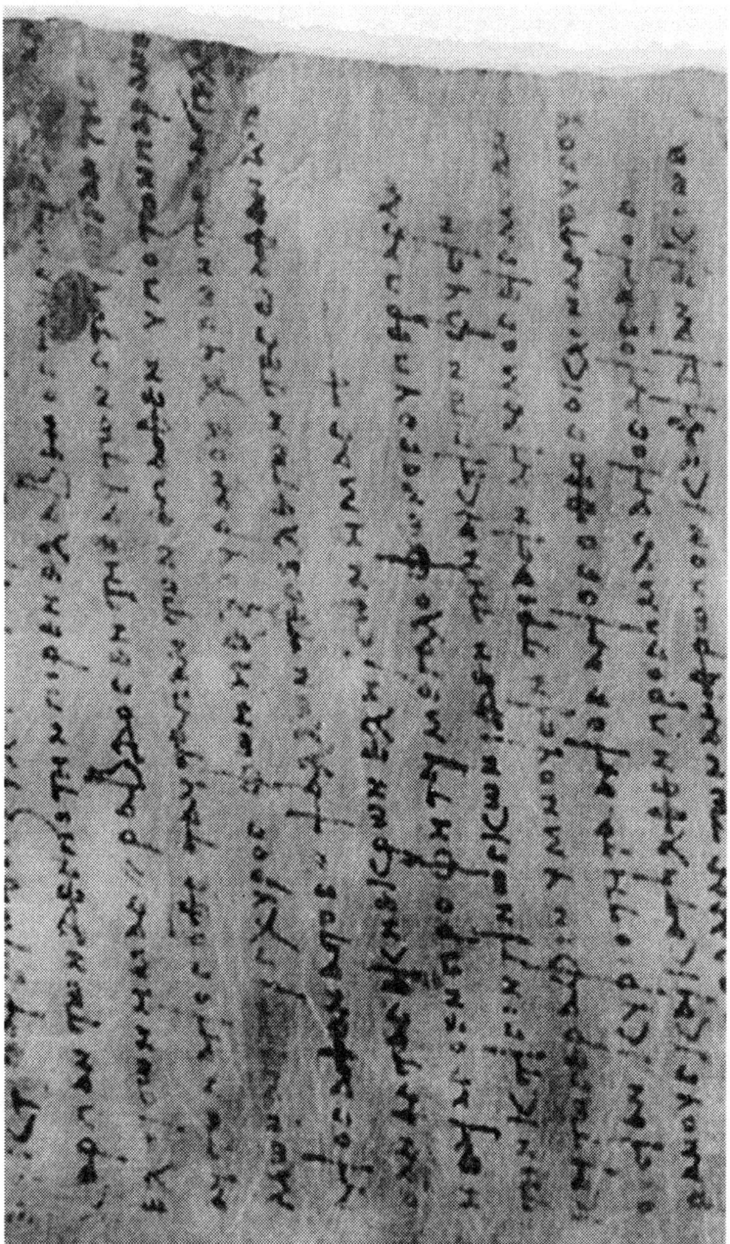

καρπαντωνδεϲποτηνπιρενελαβενοϲταυρωθιϲδιη̣μαϲ
εληιϲωνημαϲ//ραβδοϲεντηϲαυτωνϲταυρωϲαντηϲ
αυτωναγιοϲοθ̅ϲ̅ ταυταπαντωνεπαθεν υποτωνπαρανο
12 μωναγιοϲιϲχυροϲ φωνηεξουρανουχυρωντεϲεαγγελο̣ι̣.
αγιοϲαθανατοϲ//ψαλλωντεϲ⟩λεγωντεϲωαθανατοϲ
οαναϲταϲεκνεκρωνεληϊϲωνημαϲ†
+ηϲαιαϲοενπροφηταιμεγαλοφωνοϲουπερπαϲαν
16 τηνκτιϲινγινωϲκωνιδεντηνακτιϲτονφυϲιν
καιτωϲεραφινυμνουϲιντριαϲιναγιαϲμοϲειϲμιαν
ουϲιανκυριοτηταagιοϲαγιοϲοθεοϲοκλιναϲτουϲου
ρανουϲκαικατηλθενπροϲημαϲαγιοϲαγιοϲαγιοϲ

---

9 η of ημαϲ partially washed away 14 space after cross of 7/8 lett.

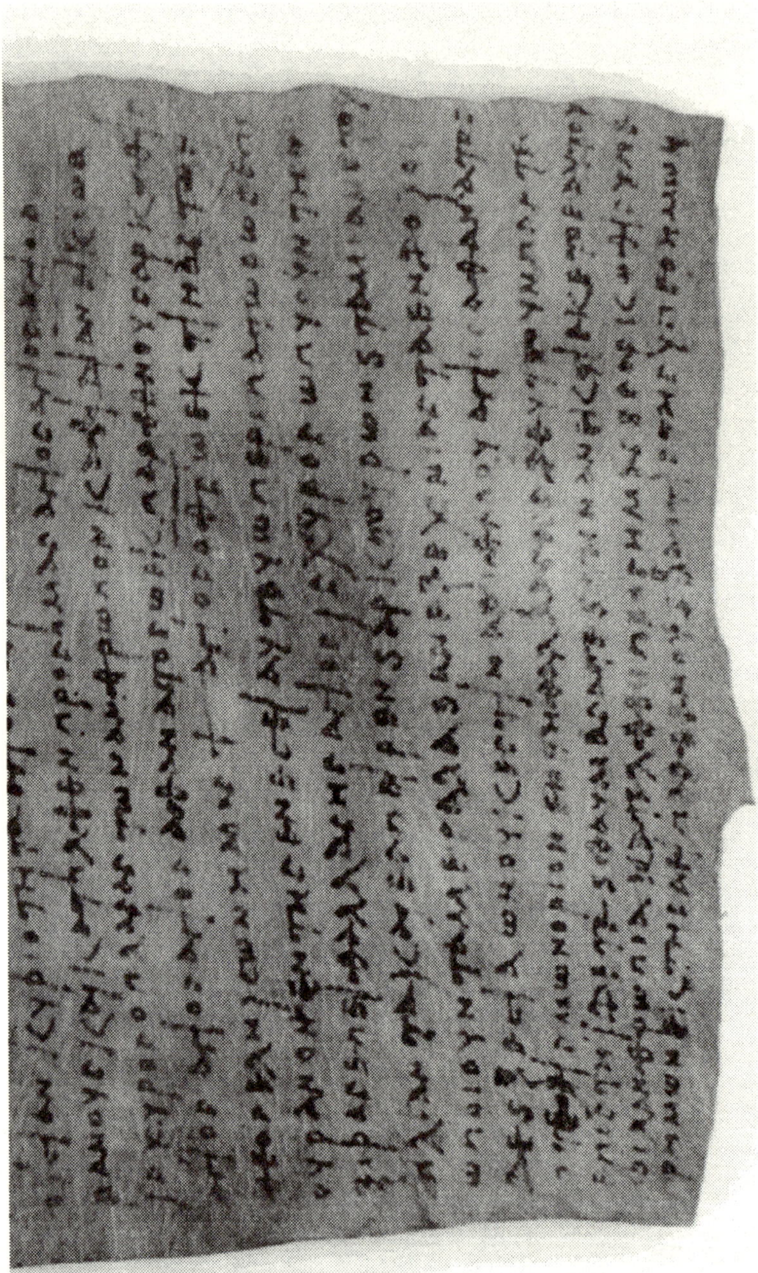

20 24 28

20 ιcχυροcοπλαcαcτωνανθρωπονκαθιδιανεικονα
αγιοcαγιοcαγιοcαθανατοcωεκπαρθενουcαρκωθιc
θεοcεληιcωνημαc† αγιοcοθ̅c̅ωεκτιναcτων
ουρανονεντηcενεcειαυτουωπεριπατωρωcεπι
24 ξιραcεπειθαλλαcηcαγιοcιcχυροcωπυουντην
πλιαντακαιεcπειρενʃαρκτουρωνʃταμιανοτου
ωποιουνταμεγαλαʃανεξεχνιαcταενδοξοι
δεʃξεcιαωνουκεcτιναριθμουαγιοcαθανατοc
28 οτεθιcμωνοριονεντηθαλλαcηcδευτεουνπαντεc
επιcτηϊδετεʃθαυμαcατεʃτηνανεκτιεκετοεαυτου
φιλανθρωπιακατελθενπροcημαcʃcαρκωθιcυπε
ρημωνεκτηcαειπαρθενουʃβαπτιcτηcυπερημων

---

22 Line over *nomen sacrum* begins over ο and extends partially over following ω

## Edited/Metrical Text

| | Syllables[93] | Stresses |
|---|---|---|
| **ἀ**cτὴρ ἐξ οὐρανοῦ | 6 | 2 |
| **β**αcιλέα cημαίνει. | 7 | 2 |
| ἅγιοc ὁ θ(εό)c, | 6 | 2 |
| **γ**εννᾶται Ἰηcοῦ⟨c⟩ | 6 | 2 |
| **δ**ιὰ λόγου θεοῦ | 6 | 2 |
| ἅγιοc ἰcχυρόc, | 6 | 2 |
| **ἐ**κ παρθένου Μαρία⟨c⟩, | 7(6) | 2 |
| **ζ**ωὴν δωρούμενοc. | 6 | 2 |
| ἅγιοc ἀθάνατοc, | 7 | 2 |
| **Ἡ**ρῴδηc ἐcπούδαcεν | 7 | 2 |
| **θ**εὸν ἀποκτείνειν. | 6 | 2 |
| ὁ ἀναcτὰc ἐκ τῶν νεκρῶν | 8 | 2 |
| ἐλέηcο⟨ν⟩ `ἡμᾶc´. | 6 | 2 |
| | | |
| **Ἰ**ωάννηc ἐβάπτιcεν | 7 | 2 |
| **κ**ύριον τὸν θεὸν ἡμῶν. | 7 | 2 |
| ἅγιοc ὁ θ(εό)c, | 6 | 2 |
| **λ**αοὶ cυνήχθηcαν | 6 | 2 |
| **μ**αρτύρων καὶ προφητῶν. | 7 | 2 |
| ἅγιοc ἰcχυρόc, | 6 | 2 |
| **ν**υκτὶ παρεδόθη | 6 | 2 |
| **ξ**ύλῳ ὃ⟨ν ἐ⟩cταύρωcαν. | 7(6)[94] | 2 |
| ἅγιοc ἀθάνατοc, | 7 | 2 |
| **ὁ** γὰρ πάντων δεcπότηc | 7 | 2 |
| **π**εῖραν ἔλαβεν. | 5 | 2 |
| ὁ cταυρωθεὶc δι' ἡμᾶc | 7 | 2 |

[93] Words such as Ἰηcοῦc 4, Μαρίαc 7 and κύριον 15 may be counted as di- or trisyllabic (see Römer p. 71). I count as trisyllabic Ἰωάννηc 14, ταμιεῖα 60 (ms. ταμια), and every instance of ἅγιοc.

[94] Depending on whether one counts the supplied augment (see ad loc.).

## Translation

A star from heaven
heralds the king.
Holy is God,
4 Jesus is begotten
through the word of God
holy and mighty,
of the virgin Mary,
8 (Jesus) the giver of life.
holy and immortal,
Herod sought
to kill God.
12 you who rose from the dead,
have mercy on us.

John baptized
the Lord our God.
16 Holy is God,
Hosts of martyrs
and prophets assembled.
holy and mighty,
20 By night he was betrayed
whom they crucified on the tree.
holy and immortal,
For the Lord of all
24 submitted to suffering.
you who were crucified for us,

| | | |
|---|---|---|
| ἐλέησον ἡμᾶς. | 6 | 2 |
| | | |
| **ῥ**αβδίσαντες αὐτόν, | 6 | 2 |
| **σ**ταυρώσαντες αὐτόν, | 6 | 2 |
| ἅγιος ὁ θ(εό)ς, | 6 | 2 |
| **τ**αῦτα πάντ⟨α⟩ ἔπαθεν | 7 | 3 |
| **ὑ**πὸ τῶν παρανόμων. | 7 | 1 |
| ἅγιος ἰσχυρός, | 6 | 2 |
| **φ**ωνὴ ἐξ οὐρανοῦ· | 6 | 2 |
| **χ**αίροντες οἱ ἄγγελοι, | 7 | 2 |
| ἅγιος ἀθάνατος, | 7 | 2 |
| **ψ**άλλοντες καὶ λέγοντες, | 7 | 2 |
| **Ὦ** ἀθάνατος | 5 | 2 |
| ὁ ἀναστὰς ἐκ νεκρῶν, | 7 | 2 |
| ἐλέησον ἡμᾶς. | 6 | 2 |
| | | |
| Ἠσαΐας ὁ ἐν προφήται⟨ς⟩ μεγαλόφωνος | 14 | 3 |
| ὁ ὑπὲρ πᾶσαν τὴν κτίσιν γινώσκων | 11 | 3 |
| εἶδεν τὴν ἄκτιστον φύσιν. | 8 | 3 |
| καὶ τὰ σεραφεὶμ ὑμνοῦσιν | 8 | 2 |
| τρι{α}σὶν ἁγιασμοῖς | 6 | 2 |
| εἰς μίαν οὐσίαν ⟨καὶ⟩ κυριότητα· | 12 | 3 |
| | | |
| ἅγιος ἅγιος ⟨ἅγιος⟩ ὁ θεὸς | 12 | 4 |
| ὁ κλίνας τοὺς οὐρανοὺς | 7 | 2 |
| καὶ κατῆλθεν πρὸς ἡμᾶς· | 7 | 2 |
| ἅγιος ἅγιος ἅγιος ἰσχυρὸς | 12 | 4 |
| ὁ πλάσας τὸν ἄνθρωπον | 7 | 2 |
| κατ᾽ ἰδίαν εἰκόνα· | 7 | 2 |
| ἅγιος ἅγιος ἅγιος ἀθάνατος | 13 | 4 |
| ὁ ἐκ παρθένου σαρκωθεὶς θεός, | 10 | 3 |
| ἐλέησον ἡμᾶς. | 6 | 2 |
| | | |
| ἅγιος ὁ θ(εὸ)ς | 6 | 2 |

have mercy on us.

Beating him,
crucifying him,
Holy is God,
all this he suffered
at the hands of lawless men.
holy and mighty,
A voice from heaven!
The angels rejoicing,
holy and immortal,
praising and proclaiming,
O Immortal
who rose from the dead,
have mercy on us.

Isaiah, the mighty voiced among the prophets,
whose knowledge transcends all the created order,
saw the uncreated nature.
And the seraphim sing
with three blessings
to one being and lordship,

Holy holy <holy> is God,
who bowed the heavens
and came down to us;
holy holy holy and mighty,
who formed man
after his own image;
holy holy holy and immortal,
God made flesh of the virgin,
have mercy on us.

Holy is God

| | | |
|---|---|---|
| "ὁ ἐκτείναϲ τὸν οὐρανὸν | | |
| ἐν τῆ̣{ϲ} ϲυνέϲει αὐτοῦ· | 15 | 4 |
| ὁ περιπατῶν ὡϲ ἐπὶ ξηρᾶϲ | | |
| ἐπὶ θαλ{λ}άϲ⟨ϲ⟩ηϲ· | 15 | 3 |
| ἅγιοϲ ἰϲχυρὸϲ | 6 | 2 |
| ὁ ποιῶν τὴν Πλειά{ν}δα | | |
| καὶ Ἕϲπερον | 11 | 3 |
| καὶ Ἀρκτοῦρον καὶ | | |
| ταμιεῖα νότου· | 10 | 3 |
| ὁ ποιῶν τὰ μεγάλα καὶ | | |
| ἀνεξιχνίαϲτα, | 14 | 3 |
| ἔνδοξά τε καὶ ⟨ἐ⟩ξαίϲια, | | |
| ὧν οὐκ ἔϲτιν ἀριθμόϲ." | 16 | 4 |
| ἅγιοϲ ἀθάνατοϲ | 7 | 2 |
| "ὁ τιθεὶϲ ⟨ἄμ⟩μον ὅριον | | |
| ἐν τῆ̣ θαλ{λ}άϲ⟨ϲ⟩η̣{ϲ}." | 13 | 4 |
| | | |
| δεῦτε οὖν πάντεϲ οἱ πιϲτοί, | 8 | 3 |
| ἴδετε καὶ θαυμάϲατε καὶ ⟨ ⟩ | 9(+) | 2/3 |
| τὴν ἀνεκδιήγητο⟨ν⟩ ἑαυτοῦ | | |
| φιλανθρωπία⟨ν⟩. | 15 | 3 |
| κατῆλθεν πρὸϲ ἡμᾶϲ, | 6 | 2 |
| καὶ ϲαρκωθεὶϲ ὑπὲρ ἡμῶν | | |
| ἐκ τῆϲ ἀειπαρθένου | 15 | 3 |
| καὶ βαπτιϲθεὶϲ ὑπὲρ ἡμῶν. | 8 | 2 |

56 "who stretched out the heaven
by his wisdom,
who walks on the sea
as on dry land;
holy and mighty,
the maker of the Pleiades
and Hesperos
60 and Arktouros and
the chambers of the south,
who does things great and
unsearchable,
glorious and amazing,
of which there is no number";
holy and immortal,
64 "who makes sand the boundary
for the sea."

Come then, all ye faithful,
behold and marvel and < >
his unspeakable love.
68 He came down to us,
made flesh for us of the ever-virgin
and baptized for us.

## III. COMMENTARY[95]

### Hymn I, 1-39/2-14

This section consists of an acrostic meditation on the central aspects of the life and passion of Christ, interlaced with three repetitions of the Trisagion refrain (see above p. 1). The birth narrative, which occupies the first Trisagion cycle (1-13), is based on that of NT Mat. 1-2. The other two cycles (14-26, 27-39) treat the baptism, passion, and ascension of Jesus.

We certainly have here a case of *contaminatio* — a pre-existent acrostic poem with the Trisagion interposed upon it. The synthesis of the two yields a highly artificial style, with the Trisagion refrain at times interrupting the syntactic flow of the main hymn (e.g., between the couplets 4/5 and 7/8; 27/28 and 30/31; 33/34 and 36). In most expanded or "farced" Trisagia one encounters the refrain with expansions accomplished through relative or participial clauses[96] rather than a running narrative with the Trisagion as a kind of farcing, which we have here. Such a union, however, should not surprise us. The Trisagion was so well-established a form, that it could adorn hymnic material with which it had little to do; e.g., a *chairetismos* to John the Baptist, *P. Lond. Copt.* 973 (van Haelst 932; 6th cent.; the following text is that of Koenen 46 and cf. n. 41):

---

[95]In the lemmata the first line numbers are those of the edited/metrical text, on which the commentary is based. The lines of the diplomatic text follow in parenthesis. All other references to the document are to the edited text, unless otherwise stated.

[96]E.g., ἅγιοc ὁ θεόc, ὃ⟨c⟩ πρὸc Ἰωάννην κάτω ἐν Ἰορδάνῃ ἀ⟨π⟩ῆλθεν --- ἅγιοc ἰcχυρόc, ὃν {αν}ἀπέκτειναν οἱ ἄθεοι κτλ. (MS. Insinger No. 32, 68 lines 3ff.; ed. W. Pleyte et P.A. Boeser, *Manuscrits coptes du musée d'antiquités des Pays-Bas à Leide*, 1897, 140 [van Haelst 761; cited after Schermann 222, Koenen 38]; cf. in our text lines 46-64).

ἅγιος ὁ θεός·
χαῖρε, Ἰωάννης, κῆρυξ ὀρθοδόξων
ἅγιος ἰσχυρός·
χαῖρε, κυρίου πρόδρομε.

Cf. also the first stanza of the Trisagion connected with a *chairetismos* to Mary (O. Skeat 14 [van Haelst 827], 6th/7th cent.; ed. H.C. Youtie, *TAPA* 81 [1950] 113 [*Scriptiunculae* I 227f.], cf. Koenen 45 n. 35), and an expanded Trisagion interlacing a prayer for the Nile flood (*Stud. Pal.* XV 250ab, 9th cent.; see now L. MacCoull, *JTS* 40 [1989] 129-35 [XIII in her *Coptic Perspectives on Late Antiquity*, London 1993]). In some cases it is not easy to decide when such mingling of hymnic types represents *contaminatio* or a technique of original composition.

This hymn well illustrates the penchant of Byzantine and Egyptian church poetry for juxtaposing the motifs of the glory and humility of Jesus and so establishing a polarity in his nature which enhances the theme of his greatness. This it does by exploiting the irony and at times precariousness of the human players in his drama, such as Mary who gives life to the life-giver (7-8), Herod who was so foolish as to try to kill God (10-11), and John who dared to baptize him (14-15). Thus the great *pantocrator* not only experiences the frailty of human nature and mortality (ọ γὰρ πάντων δεσπότης πεῖραν ἔλαβεν, 23-24), but he also submits himself to the vicissitudes of human action, both as beneficiary (Mary, John) and as victim (Herod, his betrayer, those who beat and crucified him). Bracketing the temporal narrative of Jesus' earthly experiences are two divine signs ἐξ οὐρανοῦ, the star at his birth (1) and the angelic φωνή heralding his ascension (33). Thus the polarities which characterize the separate events of his life, expressed at the level of the individual couplets of the hymn, are reinforced by a larger polarity which overarches this entire first section: although the divine Jesus receives benefits from righteous mortals and suffers disgrace and death at the hands of evil men, the two heavenly signs reveal who he truly is, i.e. king (2) and the conqueror of death (37f.). This general context of polarity provides important perspectives on the hymn's acrostic structure.

The frequent use of alphabetic or abecedarian acrostics in Christian liturgies and hymns[97] certainly received impetus from Jesus' assertion in Revelation 22.13, ἐγὼ τὸ ἄλφα καὶ τὸ ὦ, ὁ πρῶτοc καὶ ὁ ἔcχατοc, ἡ ἀρχὴ καὶ τὸ τέλοc[98] (see also 1.11; cf. 1.17 and 2.8; of the Father, 1.8, 21.6).[99] This statement is, among other things, an example of polar expression: the setting forth of the two extremities which incorporate the whole.[100] Cf., e.g., from the vespers office of the Mozarabic liturgy: *Alpha et Omega, initium et finis, deus et homo, infinitus et praefinitus, in quo et principium deitas, et ultimum sentitur humanitas; excedens omnia, vivificans cuncta, et continens universa, misere nobis* etc. (Migne PL 86.176; cf. F. Cabrol, *DACL* I 24 s.v. ΑΩ).[101] With the "alpha and omega" self-

[97]For general bibliography see C. Römer, intro. to *P.Köln* IV 172 (p. 38); K. Treu, J. Diethart on MPER N.S. XVII 56, p. 103.

[98]Similarly, a pagan hymn to the demiurge, *P.Köln* I 6.4 (= ed. princ. L. Koenen, J. Kramer, *ZPE* 4, 1969, 19-20 and repr. in *New Docs* I p. 66ff.): χαῖρ'] ἀρχή, χαῖρε τελευτ̣ή. For other pagan parallels of the ἀρχή/τέλοc (*vel sim.*) polarity, expressing a god's all-encompassing δ̣ύναμιc, cf. O. Weinreich, *ARW* 19 (1916-19) 181ff.

[99]That Revelation designates both Jesus and Yahweh as "Α and Ω, the first and last," made this ascription very adaptable to the concept of *homoousia* and a valuable weapon of orthodoxy against Arianism. This fact partly accounts for its widespread use in Christian iconography; see Dornseiff, *Alphabet* 123f. For its iconographic use in various Christian monuments, see in general the indices in Schiller I 199 and III 256 s.v. alpha und omega (Eng. trans. II 662); H. Leclercq, *DACL* I 1-24 s.v. ΑΩ; more specifically, for an early example (later 4th cent.), see the *pantocrator* of the catacomb of Commodilla (Rome) in *EEC* II 1024 fig. 169; A. Grabar, *Christian Iconography: A Study of Its Origins* (Princeton 1968) p. 34, fig. 81.

[100]Dornseiff, *Alphabet* 185; in general see E. Kemmer, *Die polare Ausdruckweise in der griechischen Literatur* (Beiträge zur historischen Syntax der griechischen Sprache 15, ed. M. Schanz); for other lit. on polarity, see Martinez' n. on *P.Mich.* XVI 757.6-10, p. 53.

[101]Among the parallels in Judaism is the Kabbalistic reading of Genesis 1.1, which understood the accusative particle את before "the heavens and the earth"

predication, Jesus declares himself the *pantocrator*, the one who encompasses and incorporates the kosmos, i.e., the orderly synthesis of all of the στοιχεῖα, represented by the alphabet.[102] ΑΩ has a similar significance in the Greek magical papyri (R.W. Daniel, *ZPE* 50, 1983, 152f.), especially in its association with the name Abrasax, whose isopsephic value is 365, thus, cosmic totality; e.g., *PGM* VII 519f., ὁ τῶν ὅλων δεσπότης, ἅγιε κανθαρε: αω· σαρθρεν Ἀβρασαξ.[103] Cf. also a Manichean Parthian hymn which also has a

---

in a manner similar to ΑΩ, as a surrogate for the entire alphabet and thus the *kosmos noetos*, the world of ideas (see Dornseiff, *Alphabet* 125). In the mystical "athbash" writing of the Hebrew alphabet (i.e., אתבש, the paring of the first and last letters, the second and second-to-last, etc.) the first pair is את = ΑΩ, the final pair is כל = τὸ πᾶν (G. Kittel in *TDNT* I 2 s.v. ΑΩ; Dornseiff loc. cit.; cf. Wm. Brashear, "The Greek Magical Papyri: an Introduction and Survey," ANRW II 18.5 [1995] 3433).♦ There was a conception that the Hebrew אמת, "truth," was the seal of God, consisting of the first, middle, and last letters of the alphabet. In Jewish literature this tradition dates to the third century AD, but it is likely earlier, since the second-cent. Gnostic Markos seems dependent on it. He portrayed ἀλήθεια as the great aeon-God, which he configured on the basis of the "athbash" pattern, αω being his head, βψ the neck, etc. (Kittel op. cit. 2f.; F. Boll, *Sphaera* [Leipzig 1903, repr. Hildesheim 1967] 471; Dornseiff, *Alphabet* 132f.).

102Dornseiff, *Alphabet* 122f. Similar is the belief, consisting of both Greek and Egyptian elements, that the 24 letters of the alphabet corresponded to the 24 hours of the day and the deities which presided over them, and thus the alphabet collectively represents the fullness of godhead (R. Reitzenstein, *Poimandres* [Leipzig 1904] 268f.; Th. Hopfner, *Griechisch—ägyptischer Offenbaurungszauber* I [Studien zur Palaeographie und Papyruskunde XXI, 1921; repr. Amsterdam 1974] § 411). Without reference to the alphabet, Paul expresses similar ideas in Colossians with regard to Jesus through the term *pleroma*: ἐν αὐτῷ εὐδόκησεν πᾶν τὸ πλήρωμα κατοικῆσαι (1.19); ἐν αὐτῷ κατοικεῖ πᾶν τὸ πλήρωμα τῆς θεότητος σωματικῶς (2.9); cf. *PGM* P 13.1f. (IV/V), [ἐ]πικαλοῦμαί σε [θεὸν τῶ]ν οὐρανῶν καὶ θεὸν τῆς γῆς καὶ θ[εὸν] τῶν διὰ [αἵματός σου] ἁγίων, τὸ πλήρωμα τοῦ αἰῶνο[ς ἡμῖν] χωρούμενο⟨ν⟩.

103Cf. V 363, 367; XIXa 47 (right). Also in Christian amulets, cf. *PGM* P 5a.16, which consists (among other things) of a cross flanked by Α and Ω, with Αβρασαξ written beneath; cf. a large ivory ring noted by H. Leclercq (*DACL* I 5, 23, s.v. ΑΩ), bearing the chrismon flanked by Α and Ω, with the legend Ἀβρασαξ. ΑΩ is not only the first and last letters of the alphabet but also, possibly more important for magic, the first and last of the traditional listing of the seven vowels, αεηιουω, which also in various configurations in magical texts

abecedarian structure, in which the gods, aeons, and light-gods offer their praises to the supreme Father of Greatness, who is called"the first *Alif* and the last *Tau*" (the first and last letters in the Syrian alphabet).[104]

The alphabet enveloping a hymn to Christ conveys the same idea longhand as the shorthand AΩ. Specifically in the Michigan Trisagion, the inscription at the top, IC XC, with the alphabetic structure that follows,[105] frames the first hymn with a kind of secret confessional statement, i.e., Ἰ(ηcοῦ)c Χ(ριcτό)c (ἐcτιν) Α-Ω. On a more

---

signifies το πᾶν, in that each vowel stands for one of the seven planets (see Martinez on *P. Mich.* XVI §§ B, G, and H, p. 110). In addition, the seven vowels may collectively designate the supreme solar god: εἴcελθε, φάνηθί μοι, κύριε, ὅτι ἐπικαλοῦμαι, ὡc ἐπικαλοῦνταί cε οἱ τρεῖc κυνοκέφαλοι, οἵτινεc --- ὀνομάζουcίν cου τὸ ἅγιον ὄνομα α εε ηηη ιιιι ooooo υυυυυυ ωωωωωωω, *PGM* IV 1002ff. (see W. Fauth, *Helios Megistos* [Leiden 1995] 112); this demiurge is later called ὁ ἐπὶ τῆc τοῦ κόcμου κεφαλῆc καθήμενοc καὶ κρίνων τὰ πάντα, ibid., 1012f. For αω with other vowel configurations in magic, see, e.g., *PGM* IV 411 (left), 433 (right), 1224; *P. Mich.* XVI § B.1, § H.14 (p. 23; *Suppl. Mag.* I 48); αω αεηιου, A. Delatte, P. Derchain, *Les Intailles magiques greco-égyptiennes* (Paris 1964) # 137(R). Cf. ωα, H. Philipp, *Mira et Magica* (Mainz am Rhein 1986) # 188 (with n. p. 116); ωωω ααα, Delatte and Derchain, op. cit. 23(R); cf. ααα ωωω, "the name of the immortal one," *Pistis Sophia* I 62 (ed. C. Schmidt, trans. V. MacDermot, Nag Hammadi Studies IX [Leiden 1978] p. 126,4). For ααα ωωω, in Christianity, cf. the inscription of the 5th cent. baptistery at Albenga (Schiller I p. 141 [Eng. trans. I 131] with plate 344). Alchemical texts associate αω or ωα with the Agathodaimon/ οὐροβόροc figure; e.g., Olyp. Alchem. *de art. sacr.* 18, M. Berthelot, *Collection des anciens alchimistes grecs* (Paris 1888) III p. 79f. (cf. Dornseiff, *Alphabet* 125).

[104]The hymn is preserved on two larger fragments and three smaller ones; see M. Boyce, *Bulletin of the School of Oriental and African Studies* 14 (1952) 437-40 (in this article she discusses other abecedarian Parthian hymns); cf. eadem, *A Reader* (see above p. 4 n. 12) p. 94 (text **ak**) and for the abecedarian material in general p. 13; H.-J. Klimkeit, *Hymnen und Gebete der Religion des Lichts* (Abhandlungen der Rheinisch-Westfälischen Akademie der Wissenschaften 79, Opladen 1989) 62. For the difficulties in dating the Middle Persian and Parthian documents, see Boyce, *A Reader* p. 14.

[105]The titular IC XC does, however, refer to the other two hymns as well; see above p. 22.

thematic level it further establishes the same ironic polarity described above, that the great *pantocrator* was born of woman and suffered shame and death.

For the Egyptian and Byzantine hymnist, the standard distinction between the "abecedarian" acrostic and the "name" acrostic loses relevance.[106] In the case of Jesus the AΩ, the entire alphabet is another one of his names of power: τὸ τοῦ cωτῆροc δὲ ῥητὸν ὄνομα Ἰηcοῦc γραμμάτων ὑπάρχειν ἕξ, τὸ δὲ ἄρρητον αὐτοῦ γραμμάτων εἰκοcιτεccάρων (Markos apud Epiph. *pan.* II [GCS 31] 18.10ff.).

**1-2 (2) ἀcτὴρ ἐξ οὐρανοῦ | βαcιλέα cημαίνει:** ms. cυμενει (η ⟩ υ Gignac I 264f.). The presents cημαίνει and the subsequent γεννᾶται, followed by a string of aorists, lend a timeless quality to these two opening strophes (Mayser II 1.131; Mitsakis § 260; see also below on ὑμοῦcιν line 43). Cf. what is sometimes called the "annalistic" pres. (*praesens tabulare*), esp. common with words such as γεννᾶcθαι, γίγνεcθαι, τελευτᾶν, etc. (KG 1.134; B. Gildersleeve, *Syntax of Classical Greek* [New York 1901, 1911] § 201; esp. J. Wackernagel, *Vorlesungen über Syntax* I [Basel 1926²] 164f.).

For the star which led the magi to Jesus (Mat. 2.2, 9-10), cf. ἀcτὴρ δὲ τοῦτον κηρύττει ἄνωθεν, S. Jos. Hymnog. I 3 (Pitra 382); cf. [Rom.] 29.77 (Pitra 237); ὃ⟨ν⟩ ἀcτὴρ καταλάμψαc ἐμήνυcεν, Maas *FK* I 5.9; πῶc μηνύcει ἀcτὴρ cε ἀcτρολόγοιc ἐρευνῶcι παιδίον νέον, S. Jos. Hymnog. I 7 (Pitra 383), cf. [Rom.] 29.93 (Pitra 240); βλέπουcα ἡ ἀμώμητοc (scil. Μαρία) --- ἀcτέρα δηλοῦντα, Rom. *cant.* Ικβʹ 1-3 (p. 8); ἀcτὴρ δὲ τοῦτον Μάγοιc οὐρανόθεν ὑπέδειξε, [Rom.] 29.69 (Pitra

---

[106]For this distinction, cf., e.g., A. Kurfess, Th. Klauser, in *RAC* I 235-38 s.v. Akrostichis. E. Peretto prefers to designate hymns and poems such as ours as "alphabetical poems," reserving the description "acrostic" for those hymns or poems whose first letters form a name or word (*EEC* I 26 s.v. "Alphabetical Poem").

236).[107] Our cημαίνει (for which I find no parallels in hymns) means that the star acts as the cημεῖον of Jesus' birth, i.e., it does not merely "reveal," him but it also "heralds" or "proclaims" prophetic fulfillment. Cf. its usage in the astrological controversy, εἰ τὰ ἄcτρα ποιεῖ ἢ cημαίνει: e.g., Clem. Alex. *eclog. proph.* 55.2 (III 152, 18f. Staehlin): οὐδ' ὄναρ τὰc ἐνεργείαc ποιεῖ τὰ ἄcτρα, cημαίνει δὲ "τά τ' ἐόντα τά τ' ἐccόμενα πρό τ' ἐόντα" (cf. E. Pfeiffer, *Studien zum antiken Sternglauben* [Leipzig/Berlin 1916] passim, but esp. pp. 72-76).

**4-5, 7 (3-4) γεννᾶται 'Ιηcοῦ⟨c⟩ | διὰ λόγου θεοῦ, | ἐκ παρθένου Μαρία⟨c⟩:** For omission of final c, cf. Gignac I 124f.; Threatte, *Grammar* I 639-40. Due to the smearing of the ink caused by water damage, the δ of διά appears filled in, as do other letters in various parts of the text (see scanned images [pp. 32ff.] and plate). On the tense of γεννᾶται see above on lines 1-2 cημαίνει. For the name "Jesus" as a designation of humanity, see below on line 8 ζωὴν δωρούμενοc.

To describe Jesus as "begotten through the word of God" seems theologically strange, since in orthodox conception he himself was the eternally existent λόγοc who was with God and who was God (Jn. 1.1), through whom God the Father created the kosmos (πάντα δι' αὐτοῦ ἐγένετο, καὶ χωρὶc αὐτοῦ ἐγένετο οὐδὲ ἕν, Jn. 1.3; cf. 1Cor. 8.6; Col. 1.16) and redeemed it from corruption (δι' αὐτοῦ ἀποκαταλάξαι τὰ πάντα εἰc αὐτόν, εἰρηνοποιήcαc διὰ τοῦ αἵματοc τοῦ cταυροῦ αὐτοῦ, Col. 1.20; cf. Jn. 1.17, 3.17; Acts 10.36; 2Cor. 5.18). The philosophic concept which the early church adopted (and indeed helped to establish through its logos-doctrine), expressed by διά with the genitive, is that of the λόγοc as ὄργανον (cκιὰ θεοῦ δὲ ὁ λόγοc αὐτοῦ ἐcτιν, ᾧ καθάπερ ὀργάνῳ προcχρηcάμενοc ἐκοcμοποίει, Philo, *leg. al.* 3.96; εὑρήcειc γὰρ αἴτιον μὲν αὐτοῦ τὸν θεὸν ὑφ' οὗ γέγονεν, ὕλην δὲ τὰ τέccαρα cτοιχεῖα ἐξ ὧν cυνεκράθη, ὄργανον δὲ λόγον θεοῦ δι' οὗ κατεcκευάcθη, idem *de cher.* 127; in the latter, note the difference between ὑφ' οὗ [agent] and δι' οὗ [instrument]. In general see

---

[107] On this hymn see Maas/Trypanis, *cant. dub.* p. xiii.

Oepke in *TDNT* II 66ff. s.v. διά; Leisegang in *RE* XIII 1.1075 s.v. logos; esp. W. Theiler, *Die Vorbereitung des Neuplatonismus* [Berlin/Zürich 1964] 27ff; 41; more recently, D.T. Runia, *Philo of Alexandria and the Timaeus of Plato* [Leiden 1986] 171-74; 447f.). Our διὰ λόγου θεοῦ may be taken as a version of this idea. Coupled with the following ἐκ παρθένου Μαρία⟨c⟩, the language approximates that of the creeds, where, however, πνεῦμα replaces λόγος.

Many credal formulae portray Christ as begotten of the Holy Spirit and the Virgin Mary: τὸν γεννηθέντα ἐκ πνεύματος ἁγίου καὶ Μαρίας τῆς παρθένου, *Rom. Symb.* (Lietzmann, *Symbolst.* 15/203). Other traditions, however, seek more precisely to delineate and differentiate the roles of Mary and the Spirit with ἐκ and διά respectively; e.g., γεννηθέντα τελείως ἐκ Μαρίας τῆς ἀειπαρθένου διὰ πνεύματος ἁγίου, [Ath.] *interpr. symb.* in *ACO* 1.1.7, p. 66,17; cf. *Epiph.* II (Lietzmann, loc. cit.; see below); cf. the (reconstructed) baptismal creed from the *Traditio* of Hippolytos (Lietzmann, *Symbolst.* 78; J.N.D. Kelly, *Early Christian Creeds* [Oxford $1960^2$] 90f.). So also in numerous formulations in Patristic literature: ἥ τε ἐκ παρθένου διὰ πνεύματος ἁγίου κατὰ σάρκα γέννησις, (Ath. *exp. Ps.* 9 [Migne PG 27.84b]); εὐδοκίᾳ τοῦ πατρὸς καὶ συνεργίᾳ τοῦ ἁγίου πνεύματος ἀσπόρως συλληφθεὶς ἀφθόρως ἐκ τῆς ἁγίας παρθένου καὶ θεοτόκου Μαρίας γεγέννηται διὰ πνεύματος ἁγίου (Jo. D. *f.o.* 2.22); cf. the dictum of Epiphanius, ἀληθινῶς ἐγεννήθη Χριστὸς ἀπὸ Μαρίας τῆς ἀειπαρθένου διὰ πνεύματος ἁγίου, οὐκ ἀπὸ σπέρματος ἀνδρός (*pan.* III [GCS 37] 515.24f.; cf. II [GCS 31] 394.16; III [GCS 37] 467.4; *anaph. gr.*, ed. G. Garittc *Museon* 73 [1960] 299, line 2). The use of διά in this credal formulation refines the theological perspective in two directions: first, since it is a preposition usually connected with creation rather than procreation, it shuns any notion of a *hieros gamos* (see Lk. 1.35 [cited below], where the spirit broods over the womb of the virgin as he did over the chaos in creation, and cf. Just. *1 apol.* 33.6, καὶ τοῦτο [scil. τὸ πνεῦμα] ἐλθὸν ἐπὶ τὴν παρθένον καὶ ἐπισκιάσαν οὐ διὰ συνουσίας ἀλλὰ διὰ δυνάμεως ἐγκύμονα κατέστησε). Second, since διά conveys secondary activity, it avoids portraying Jesus as the child of the Spirit, but rather of the Father, who begets through the Spirit (for the problem in general, see Aug., *enchirid.* 37-40 [Migne, PL 40.251f.]; Kelly, op. cit. 377).

Significantly, with regard to our διὰ λόγου θεοῦ, Biblical and Patristic thought closely associated or identified the Holy Spirit with the logos, e.g., in the areas of creation (λόγοc – πνεῦμα/cοφία in LXX Ps. 32.6; Prov. 3.19-20; 8.22-25; Gen. 1.2; cf. Iren. *haer.* 4.20.3 [Migne PG 7.1033c]; Thphl. Ant. *Autol.* I 7 and esp. II 10 [Migne PG 6.1036a, 1064b-1065b]; Ath. *ep. Serap.* 1.31 [Migne PG 26.601a]), prophetic inspiration (for the Bible, R. Brown, *The Birth of the Messiah* [New York 1993[2]] 274; cf. τὸ προφητικὸν πνεῦμα in Just. *1 apol.* 31.1, 39.1 [Migne PG 6.376a, 388b] with θεῖοc λόγοc *ibid.* 33.9, 36.1 [Migne PG 6.381c, 385a]; see esp. Ath. *ep. Serap.* 1.31 [Migne PG 26.601bff.]; Thphl. Ant. *Autol.* II 10), and the incarnation. With respect to the last, especially important is the aforementioned Lk. 1.35, where Gabriel tells Mary the manner by which Jesus will be conceived in her womb: πνεῦμα ἅγιον ἐπελεύcεται ἐπὶ cὲ καὶ δύναμιc ὑψίcτου ἐπιcκιάcει coι. Athanasius understands πνεῦμα and δύναμιc as the Holy Spirit and the logos (or Son) respectively: οὕτω καὶ επὶ τὴν ἁγίαν παρθένον Μαρίαν ἐπιδημοῦντοc τοῦ λόγου, cυνειcῆρχετο τὸ πνεῦμα, καὶ λόγοc ἐν τῷ πνεύματι ἔπλαττε καὶ ἥρμοζεν ἑαυτῷ τὸ cῶμα (Ath. *ep. Serap.* 1.31[Migne PG 26.605a]). Justin attributes both to the λόγοc· τὸ πνεῦμα οὖν καὶ τὴν δύναμιν τὴν παρὰ τοῦ θεοῦ οὐδὲν ἄλλο νοῆcαι θέμιc ἢ τὸν λόγον (*1 apol.* 33.6 [Migne PG 6.381b]). With both perspectives cf. Tertul. *adv. Prax.* 26.4 (CC Tertul. 2.1196 [Migne PL 2.189a]): *Hic Spiritus Dei idem erit sermo. Sicut enim Iohanne dicente: "Sermo caro factus est," Spiritum quoque intellegimus in mentione sermonis, ita et hic sermonem quoque agnoscimus in nomine Spiritus. Nam et Spiritus substantia est sermonis et sermo operatio Spiritus et duo unum sunt.*

The λόγοc in our text may thus be understood as the pre-incarnate Son acting as the instrument of his own conception or as the Son in union with the Spirit, reflecting the Athanasian interpretation of Lk. 1.35 and the common credal formula. Cf. also Tertul. *adv. Jud.*

13.22 (CC Tertul. 2.1389 [Migne PL 2.636c]); Iren. *haer.* 5.1.3 (SC 153 p. 27f.); Gr. Nyss. *Apoll.* III 1.191.22ff.; Cyprian *de idol. van.*, Migne PL 4.599 and cf. n. 69). For specific authors, see E.R. Goodenough, *The Theology of Justin Martyr* (Jena 1923) 181f.; L.W. Barnard, *Justin Martyr* (Cambridge 1967) 103f.; A.W.F. Blunt, *The Apologies of Justin Martyr* (Cambridge 1911) intro. xxviii; C.R.B. Shapland (trans.), *The Letters of Saint Athanasius concerning the Holy Spirit* (London 1951) 145f. and esp. n. 26. More generally, J. Meyendorff, *Byzantine Theology* (New York 1983[2]) 170f.; Kelly, op. cit. (see above p. 51) 148f. The fact, however, that we may explain γεννᾶται Ἰηcοῦc --- διὰ λόγου θεοῦ from these well-known motifs in no way mitigates the significance that the idea is expressed here in this specific form for the first time.♦

**8 (4) ζωὴν δωρούμενος:** cf. ἅγιοc εἰcχυρ[ὸc ὁ cωτ]ὴρ ζωὴν δωρούμενοc καὶ θάνατον πατούμ̣[ενοc], *The Monastery of Epiphanius at Thebes* II, ed. W.E. Crum, H.G. Evelyn White (New York 1926) 132, O.598, 5f. (Koenen 40 with n. on 6ff.); δωρούμενοc τὴν ζωὴν ὁ ἀναcτὰc ἐκ τῶν νεκρῶν, *O.Bodl.* II 2166; cf. τὴν ζωὴν ἡμῖν δωρήcαc, MPER N.S. XVII 40.3f., 11f. Comparable terminology includes ζωοδότηc (*P. Köln* IV 172.10 with Römer's n. ad loc.) and ζωοποιεῖν/-ποιόc (NT Jn. 5.21, 1Cor. 15.22; *ep. Barn.* 7.2, 12.5, 7; MPER N.S. XVII 16.5, 8f.); cf. also Jn. 10.28 and 2Pet. 1.3. Whereas the description of Jesus as life-giver usually focuses on his power over death, in our text it expresses the paradox that the giver of life was begotten; cf. [Rom.] 19νθʹ, Pitra 233, μυcτήριον ἐφάνη --- κτίζεται ἄκτιcτοc --- ὁ ὢν γίνεται (see further the intro. to this section).

The last four cola (4-5, 7-8) stand together as a single sentence, interrupted by the Trisagion refrain (see above p. 44f.). Stylistically, they form a kind of chiasmus: γεννᾶται Ἰηcοῦ⟨c⟩ (A) διὰ λόγου θεοῦ (B) ἐκ παρθένου Μαρία⟨c⟩ (B) ζωὴν δωρούμενοc (A). That is, Jesus, the one given life, through the logos/pneuma and of the virgin, is the life-giver. The two descriptions of the Son (*Christus patiens, Christus agens*) bracket the divine and human forces involved in his conception (for polarity in the first hymn in general, see intro. to this section). In addition, if we take ζωὴν δωρούμενοc as designating divine power (as so many of the titles in this poem: βαcιλέα, 2; θεόν, 11; κύριον τὸν θεὸν ἡμῶν, 15; ὁ γὰρ πάντων δεcπότηc, 23) we may see the two "A" cola as an expression of the Son's two natures, since the name Ἰηcοῦc may describe his humanity and historical incarnation (ὁ θεὸc λόγοc, ἐνυπόcτατοc ὢν θεόc, μεταcχὼν ἀνθρωπότητοc, Ἰηcοῦc ἐκλήθη. --- ὁ γὰρ νῦν κληθεὶc Ἰηcοῦc διὰ τὴν ἀνθρωπότητα, θεὸc δὲ λόγοc ὢν καὶ πρὸ τῆc τοῦ ὀνόματοc κλήcεωc, καὶ μετὰ τὴν τοῦ ὀνόματοc κλῆcιν, αὐτόc ἐcτιν Ἰηcοῦc, δι' οὗ τὰ πάντα, θεὸc ὢν κατὰ πνεῦμα, ἄνθρωποc δὲ διὰ τὴν τοῦ δούλου μορφήν, [Ath.] *dial. Trin.* 5.8 [Migne PG 28.1269b]; cf. Ath. *Ar.* 1.42, Migne PG 26.100a; ibid. 2.16, Migne 26.181a; Lampe s.v. Ἰηcοῦc, C1b; Foester in *TDNT* III 287-93 s.v. Ἰηcοῦc). Thus the configuration becomes human nature—divine conception—human conception—divine nature. Such a Christology, of course, does not mesh with a rigorously Monophysite orientation, but there is no firm proof that our hymn subscribes to such, and systems of thought which have been labeled Monophysite have many theological shadings and gradations (see intro. above pp. 16-20).♦

**10-11 (4-5) Ἡρῴδηc ἐcπούδαcεν | θεὸν ἀποκτείνειν:** ms. αποκτινεν (ει 〉 ε Gignac I 257f.; but -εν as an infinitive ending is probably more an issue of morphology than phonology [Gignac II 330f., 362]). The folly of Herod's attempt to kill the divine Jesus through the "Massacre of the Innocents" is a common theme in nativity hymns: δέcποτα --- ὃν --- Ἡρῴδηc θανεῖν ἐβουλεύετο, Maas *FK* I 5.7ff.; Ἡρῴδηc ἀνελεῖν cε βουλόμενοc ὡc θεόc, Follieri II 70 (on the theme of irony, see introduction to this section).

**14-15 (6) Ἰωάννηc ἐβάπτιcεν | κύριον τὸν θεὸν ἡμῶν:** ms. Ιωαννην (cf. ms. δεcποτην below, 23-24). -c 〉 -ν Mayser/Schmoll I 1.182f., Gignac I 131f.; here perhaps due to either confusion of cases or influence of preceding ν's. In 15 the number of stresses is two; the oblique cases of ἡμεῖc, ὑμεῖc and αὐτόc may be regarded as unaccented after a word accented on the final syllable (Maas/Trypanis p. 512 D).

"The Lord God" (יהוה אלהים) or "the Lord our God" were titles of Yahweh and reflect the lofty Christology common in Byzantine/Egyptian church poetry. Such titles were often used of Christ in his baptism as evidence of his humility and *kenosis*, in that Christ, as God, submitted himself to be baptized by a man (cf. on 68-70 and 70 below, where Jesus' baptism is an aspect of *katabasis*). E.g., μέγα μυcτήριον καὶ φοβερόν, ὅτι δοῦλοc δεcπότην ἐβάπτιcεν, Maas *FK* I 5.17 (cf. Rom. *cant.* 5δ′4-5 p. 36). Similarly, in a hymn by Romanos, John the Baptist objects, comparing himself to Uzzah (2Sam. 6.6ff.), whom God struck with lightning for touching the sacred ark, νῦν δὲ κεφαλὴν κρατοῦντα με τοῦ θεοῦ μου πῶc με οὐ φλέξει; (*cant.* 5ιβ′8-9 p. 39). In general, for baptismal themes in this hymn see intro. above pp. 22-25.

**17-18 (7) λαοὶ cυνήχθηcαν | μαρτύρων καὶ προφητῶν:** ms. cυνεχθηcαν. ε- often replaces the η- augment in Byzantine period Greek either for phonological reasons (η ⟩ ε accented and after nasal, Gignac I 243; cf. Mayser/Schmoll I 1 47f.) or because of confusion with the syllabic augment (Gignac II 233f.; cf. also κατελθεν 68 (30) below). For the general neglect of temporal augment during the later periods, see Jannaris, op. cit. (above, p. 10 n. 24) § 717; K. Dieterich, *Untersuchungen zur Geschichte der griechischen Sprache* (Byzantinishes Archiv 1), Leipzig 1898, 210f.

I find no parallels for the phrase λαοὶ --- μαρτύρων καὶ προφητῶν; cf., however, a similar usage of δῆμοc: προφῆται, ἀπόcτολοι, μαρτύρων δῆμοc, Follieri III 371; δῆμοι μαρτύρων ἄπειροι, Chrys. *hom. in Rom.* 12.4 (Migne PG 60.499, 29f.); cf. idem, *hom. dict. in praes. imp.* 4 (Migne, PG 63.476, 45f.); τῶν μαρτύρων οἱ δῆμοι, Thdt. *Ps.* 46.6 (Migne PG 80. 865f.); Lampe s.v. δῆμοc 2. Cf. also Heb. 12.1 and Rev. 7.9, where the souls of the μάρτυρεc are a τοcοῦτον νέφοc and a ὄχλοc πολύc respectively. Similarly Polycarp expresses gratitude to be included ἐν ἀριθμῷ τῶν μαρτύρων (*mart. Pol.* 14.2).

μάρτυρεc and προφῆται are among the groups of the blessed who are asked to intercede on sinners' behalf, e.g., in hymns of the *Apodeipnon* (προφητῶν καὶ μαρτύρων πάντεc χοροὶ --- ὑπὲρ πάντων πρεcβεύcατε ἱλαcμόν, Maas, *FK* I 1.7f.; cf. *ibid.* 2.7-10; 3.9-12 [here, as in Folleri above, the triad προφῆται, ἀπόcτολοι καὶ ἅγιοι μάρτυρεc]) and other liturgical traditions (*lit. Copt. Jac.*, Brightman 169.6-25; from the Coptic baptismal rite, Woolley, *Coptic Offices* 49). Sometimes their souls are among the divine hosts who join in the heavenly adoration, which human worship ideally imitates: (ὑμνοῦcιν) --- πνεύματα δικαίων καὶ προφητῶν, ψυχαὶ μαρτύρων καὶ ἀποcτόλων, *lit. Jac.*, Brightman 50.21f; cf. Trypanis "Three Hymns," 3.8. For the close association of martyrs and prophets (who were often killed for preaching the word of God) in both Judaism and Christianity, cf. H.A. Fischel, *Jewish Quarterly Rev.* 37 (1947) 265-80, 363-86; cf. also Isaiah's epithet προφητομάρτυc in *de Is. proph.*, Pitra 454. Jewish/Christian martyrological traditions portray them, like Jesus, as sacrificial victims (Fischel op. cit. 374 with nn. 145 and 146; Charles on Rev. 6.9, I 172-74; Michel in *TDNT* VII 934f. s.v. cφάζω; *mart. Pol.* 14.1-2).

It is difficult to identify our προφῆται and μάρτυρεc and what role they play; indeed, these lines seem to break the flow of the narrative. The couplet may simply provide a liturgical aside or parenthesis, describing the worshipping hosts of saints, apostles, prophets, and martyrs, abbreviated to λαοὶ --- μαρτύρων καὶ προφητῶν for the acrostic and the rhythm. The use of cυνάγεcθαι, practically a *terminus technicus* for the gathering of worshippers (Lampe s.v. cυνάγω A 2; cf. cυναγωγή, Lampe s.v. C), supports this understanding, as does the possible baptismal/liturgical context of the hymn in general (see intro. above pp. 22ff.). cυνάγεcθαι, however, also has military connotations (LSJ s.v. I 3; cf. cυνηγμένοι εἰcὶν ἐφ' ἡμᾶc πάντεc οἱ βαcιλεῖc τῶν 'Αμορραίων, LXX Jos. 10.6) as does λαοί ("army" in both classical and Biblical usage; cf. Martinez in *Ancient Magic and Ritual Power*, ed. M. Meyer and P. Mirecki [Leiden 1995] 339 n. 15). Our hymn possibly portrays the company of prophets and martyrs as a mighty gathered army, strategically placed between the baptism and crucifixion, two events in which Jesus himself engages the powers of chaos and death (with regard to baptism, see below, p. 76). If this military interpretation is correct, it could provide an antithesis to the Biblical motif of the worldly ruling powers assembled against Christ, e.g., in Ps. 2.2 (quoted in Acts 4.26), οἱ ἄρχοντεϲ ϲυνήχθηϲαν ἐπὶ τὸ αὐτὸ κατὰ τοῦ κυρίου καὶ κατὰ τοῦ χριϲτοῦ αὐτοῦ· cf. Is. 43.9, πάντα τὰ ἔθνη ϲυνήχθηϲαν ἅμα, καὶ ϲυναχθήϲονται ἄρχοντεϲ ἐξ αὐτῶν.

**20-21 (8) νυκτὶ παρεδόθη | ξύλῳ ὅ⟨ν ἐ⟩ϲταύρωϲαν:** ms. παρετωθη (intervocalic δ ⟩ τ, Gignac I 82); ms. ξυλου (final ω(ι) ⟩ ου *ibid.* 209f., 213f. [where he suggests bilingual interference; cf. Teodorsson 234, 236], but this may be a case of syntactic confusion of dative and genitive [Gignac II 22 with n. 5]. For the dative with ϲταυροῦν, cf. καὶ [scil. ὁ Ϲατανᾶϲ] ϲυμβουλεύει ξύλῳ ϲταυρῶϲαι τὸν Ἰηϲοῦν, [Chrys.] *exeunt pharis.*, Migne PG 61.709.16; the normal phrase, however is ἐπὶ ξύλου; see below); ms. ο ϲταυρωϲαν (om. final -ν, Gignac I 111-14; F. Völker, *Papyrorum Graecarum Syntaxis Specimen* [Bonn 1900] 31. For the occasional omission of syllabic augment in the aor. and imperf. of simplex verbs, cf. Gignac II 223-25; in the case of ⟨ἐ⟩ϲταύρωϲαν its loss is facilitated by its unaccented position [B.G. Mandilaras, *The Verb in the Greek Non-literary Papyri*, Athens 1973, § 236; see below on 62 ⟨ἐ⟩ξαίϲια, p. 79] and the preceding ο which frequently interchanges with ε before ϲ [Gignac I 289-92]). Line 21 has two stresses, since relatives could be regarded as unaccented with respect to the rhythm (so in Romanos; see Maas/Trypanis p. 512 F).

For Jesus' betrayal by night, see Mk. 14.43-50 with parallels; 1Cor. 11.23 (institution of Eucharist), ὁ κύριοϲ Ἰηϲοῦϲ ἐν τῇ νυκτὶ ᾗ παρεδίδετο ἔλαβεν ἄρτον κτλ. The phrase "whom ______ crucified" (*vel sim.*) especially emphasizes the guilt of the Jews in the NT: κύριον αὐτὸν καὶ χριϲτὸν ἐποίηϲεν ὁ θεόϲ, τοῦτον τὸν Ἰηϲοῦν ὃν ἐϲταυρώϲατε, Acts 2.36, cf. 4.10; and with the ξύλον motif, cf. ὃν καὶ ἀνεῖλον κρεμάϲαντεϲ ἐπὶ ξύλου, Acts 10.39; ὁ θεὸϲ --- ἤγειρεν Ἰηϲοῦν ὃν ὑμεῖϲ διεχειρίϲαϲθε κρεμάϲαντεϲ ἐπὶ ξύλου, Acts 5.30. Our hymnist adapts such Biblical motifs to an acrostic and rhythmical structure, resulting in a somewhat poetic word order for the cola.

The use of ξύλον for ϲταυρόϲ is peculiar to the NT and lit. dependent on it (cf. in addition to refs. cited above, Acts 13.29; Gal.

3.13; 1Pet. 2.24). It is, however, paralleled by the secular meaning "gallows" or "stake for impalement" (Ar. *ran.* 736 with scholiast ad loc.; Alexis Com. 224.10; cf. *arbor infelix*, Cic. *Rab. Perd.* 13; Liv. I 26.6) and is based on the Jewish practice of hanging executed criminals on trees for public exposure (Deut. 21.23 [cited in Gal. 3.13]; Josh. 10.26; Jos. *ant.* 4.202; Philo *som.* 2.213). In hymns cf. ξύλῳ (sc. αὐτὸν) προϲήλωϲαν, *anepigr.* XV 1β′, Pitra 482; also ibid. 483 (ibid. δ′); Rom. *cant.* 20α′ 2f. (p. 149), 21κ′ 7 (p. 163), 22β′ 3-7 (p. 165); Follieri II 568-70 passim; etc. In general see s.v. ξύλον in Bauer/Aland and *TDNT* V 37, 39f.; F.F. Bruce, *The Acts of the Apostles* (Grand Rapids 1990$^{3}$) 172 (on 5.30). On the development of the image of ξύλον (esp. with regard to "tree") in Patristic lit., see Lampe s.v. C1; MacCulloch 334-36; J.B. Lightfoot, *The Apostolic Fathers* 2.2 (London 1889, 1890 [repr. Peabody, Mass. 1989]) p. 291 (on Ign. *Smyr.* 1.2); G.Q. Reijners, *The Terminology of the Holy Cross in Early Christian Literature* (Nijmegen 1965) 221 (index s.v. ξύλον). The more generic meaning "wood," however, is also important, given the typological correlation between the wood borne by Isaac for the altar of his own sacrifice and the wood of the cross, borne by Christ and, by extension, Christians (e.g., Iren. *haer.* IV 5.4 [SC 100 pp. 433-35]; cf. J. Gribomont in *Dictionnaire de spiritualité* VII 1993f.; R. Martin-Achard in *Anchor Bible Dictionary* III 470 s.v. Isaac).

**23-24 (8-9) ọ γὰρ πάντων δεϲπότηϲ | πεῖραν ἔλαβεν:** ms. καρ (initial γ ⟩ κ Gignac I 77); ms. δεϲποτην (final ϲ ⟩ ν cf. above 14-15 Ιωάννηϲ [ms. Ιωαννην]); ms. πιρεν (α ⟩ ε especially frequent before and after liquids and nasals; Gignac I 278-82, 285f.; Dieterich op. cit. [above p. 55], 3-11).

As a name for God, δεϲπότηϲ occasionally in the LXX translates אדון or אדוני (Gen. 15.2, 8; Is. 1.24, 3.1; Jer. 1.6; 4.10; etc.) and very rarely יהוה (Pr. 29.25); cf. E. Hatch, H.A. Redpath, *A Concordance to the Septuagint and the Other Greek Versions of the Old Testament* (Oxford 1897, 1900) s.v.; Rengstorf in *TDNT* II 45-48 s.v. In the NT it designates the Father (only as a direct address in prayer: Lk. 2.29, Acts 4.24, Rev. 6.10) and Jesus, but only in the later material (Jude 4, 2Pet. 2.1; Rengstorf op. cit. 48f.; J.N.D.

Kelly, *A Commentary on the Epistles of Peter and Jude* [New York 1969], p. 252). It is, however, frequently used of the Son in hymns (e.g., *Monastery of Epiphanius* [see above p. 53] 593.23 [p. 130]; Maas *FK* I 5.7, 17; *P. Köln* IV 173.9) and in Christian papyri (Preisigke vol. I s.v. 5).

Specifically for the title ὁ πάντων δεcπότηc *vel sim.* (not in NT; cf. Jesus as κύριοc πάντων, Acts 10.36) see LXX Job 5.8; Wisd. Sol. 6.7, 8.3; cf. Sir. 36.1; Rengstorf op. cit. 46, 47. The earliest Fathers used it chiefly of the first person of the Trinity, often joined with other ascriptions: ὁ μέγαc δημιουργὸc καὶ δεcπότηc τῶν ἁπάντων, 1Clem. 20.11; similarly 33.2; τοῦ πατρὸc τῶν πάντων καὶ δεcπότου θεοῦ, Just. *1 apol.* 46.5 (Migne PG 6. 397c); similarly ibid. 61.3, 10 (Migne PG 420c, 421b); Diogn. 8.7; cf. in Christian letters Naldini, op cit. (above p. 29 n. 85) 42.3 (IV) and 61.29f. (IV); in general, see Bauer/Aland s.v.; Lietzmann, *Symbolst.* 32-34/220-22. Later Patristic lit. and hymnic material employ it prolifically for both the Father and the Son (for the latter, cf. the last line of each stanza of Rom. *cant.* 11; further examples below). In late-antique pagan sources, cf. εἷc θεὸc ὁ πάντων δεcπότηc (inscription from Samaria [late III/IV AD], J.W. Crowfoot, et al., *The Objects from Samaria* [Samaria-Sebaste III, London 1957] 37, no. 12 [pl. V 3], republished in Horsley, *New Docs* 1, pp. 105-07). In third-/fourth-cent. AD magical texts ὁ πάντων δεcπότηc describes the great demiurge (*PGM* III 589, XII 250; δ. τοῦ παντόc IV 1164); cf. ὁ τῶν ὅλων δεcπότηc, VII 519. Thus the title in our text emphasizes deity, as do other titles of Jesus in hymn 1 (θεόc 11, κύριοc ὁ θεόc 15), and, like them, in its context it expresses the irony that the supreme God submitted to human experience, suffering, and injustice (cf. λογίcαcθε — πόcην παρὰ τῶν ἁμαρτωλῶν ἀντιλογίαν ὑπέμεινεν ὁ τῶν ἁπάντων δεcπότηc, Thdt. *Heb.*, Migne PG 82.772a; ἐcταύρωcαν αὐτὸν τὸν ἑαυτῶν καὶ πάντων δεcπότην, [Chrys.] *hom. 2 in Ps. 50*, Migne PG 55.580.1; on this point in general see above pp. 45f.).

πεῖραν λαμβάνειν/λαβεῖν, similar to the metaphorical use of γεύεcθαι (LSJ s.v. II 3), means principally "make trial of" or "(fully) experience" (cf. πεῖραν λαβὼν καὶ γευcάμενοc τῆc φιλανθρωπίας τοῦ θεοῦ, Chrys. *hom. 1-24 in Eph.* 5.18.2 [Migne, PG 62.123.18f.])." It is unusual that our text uses the words without a genitive, which almost always patterns with this idiom, whether the experience is positive or negative: τῆς ἰατρείας πεῖραν λήψῃ, Chrys. *ep. ad Olymp.* 9.4d; πεῖραν --- τῆς γνησίας φιλίας --- λαβόντες, idem *ep.* 38 (Migne PG 52.631); μαστίγων πεῖραν ἔλαβον (sc. οἱ πιστοί) "they suffered scourgings," NT Heb. 11.36; πεῖραν λαμβάνειν τῶν ἐξακολουθού⟨ν⟩των ἐπιτ[ί]μων, *UPZ* I 110.129f.; in general see C. Spicq, *Theological Lexicon of the New Testament* (trans. and ed. J.D. Ernest, Peabody Mass. 1994) III 81f. s.v.; F. Field, *Notes on the Translation of the NT* (Cambridge 1899) 232f. Occasionally, however, the genitive need not be expressed if easily inferred from the context: ἐπὶ γὰρ τῶν ἀνθρώπων ἀδύνατον εἰδέναι τὴν κάκωσιν τοῦ κακουμένου τὸν μὴ πεῖραν λαβόντα καὶ διὰ τῶν αἰσθητῶν ἐλθόντα. πάντα ὑπέστη ὁ ἀρχιερεὺς ὁ ἡμέτερος, Chrys. *hom. in Heb.* 7.2 (Migne 63. 63.35ff.); cf. idem *fr. Job* (Migne PG 64.589a), where he comments on Job 5.19 (ἑξάκις ἐξ ἀναγκῶν σε ἐξελεῖται, ἐν δὲ τῷ ἑβδόμῳ οὐ μὴ ἅψηταί σου κακόν) as follows: τουτέστιν --- ἐν μὲν τοῖς πρώτοις ἀφίησι πεῖραν λαβεῖν· μετὰ δὲ ταῦτα, οὐδὲ πεῖραν, ὥστε μηδὲ τοῦ ῥυσθῆναί σε χρείαν εἶναι λοιπόν.

πεῖραν λαβεῖν *vel sim.* has broad currency in Patristic lit. as a phrase indicating the entire range of Jesus' human experience, particularly his incarnation (αὐτὸς ἔγνω τὸ πλάσμα ἡμῶν καὶ οὕτως· πεῖραν ἔλαβε τοῦ πλάσματος ἡμῶν, Didym. *fr. Ps.* 983 [II 233.27f.]; cf. idem *in Zach.* I 250.5) and death (ὁ --- υἱὸς τοῦ θεοῦ, --- ἀθάνατος καὶ θανάτου πεῖραν λαβὼν σαρκί, [Ath.] *contr. Ar.*, Migne PG 28.464b; ἀδύνατον γὰρ θανάτου πεῖραν λαβεῖν τὸν μὴ πρότερον γεννηθέντα, Bas. *hom.* 13.1, Migne PG 31.424b; cf. Didym. *Jo.* frag. 12.4; Epiph. *pan.* III [GCS 37] 271.14; cf. also the parallel usage of γεύεσθαι in Heb. 2.9, ὅπως [sc. Ἰησοῦς] --- ὑπὲρ παντὸς γεύσηται θανάτου). With regard to our hymn and the omitted genitive, I suggest two possibilities. The πάντων of line 23 may be assumed in 24 as well; thus: "the Lord of all suffered (all)," i.e., all things relating to death and mortality. The phrase would thus closely parallel line 30 below: ταῦτα πάντ⟨α⟩ ἔπαθεν. Cf. also πάντων γὰρ ἔλαβε τῶν ἀνθρωπίνων πεῖραν, Chrys. *hom. in Heb.* 7.2 (Migne PG 63.63.2). It may, however, be preferable not to make πάντων perform double duty and simply allow the surrounding context of Jesus' experience of mortality and suffering to supply the referent of πεῖραν ἔλαβεν.

Theodoret makes the term πεῖρα a focal point of his defense of Christ's "two natures" against Monophysism, e.g., in his comment on Heb. 4.15, that not only did Jesus know the weakness of our nature as God, ἀλλὰ καὶ ὡc ἄνθρωποc πεῖραν τῶν ἡμετέρων ἔλαβε παθημάτων (Migne PG 82.708d). Cf. also apud Cyr. *apol. Thdt.* 10 (*ACO* 1.1.6, p. 136,22f., Migne PG 76.436d [on Heb. 5.8]), τίc ὁ πείρᾳ μαθὼν τὴν ὑπακοὴν καὶ ταύτην ἀγνοῶν πρὸ τῆc πείραc; --- οὐχ ὁ θεὸc λόγοc ὁ ἀθάνατοc ὁ ἀπαθὴc ὁ ἀcώματοc ---· ἀλλὰ τὸ ἐκ cπέρματοc Δαυὶδ ληφθὲν ὑπ' αὐτοῦ, τὸ θνητὸν τὸ παθητὸν τὸ δεδιὸc τὸν θάνατον ---. ἐν πείρᾳ δὲ γέγονε τῶν ἡμετέρων παθημάτων ἄνευ ἁμαρτίαc ἡ ἐξ ἡμῶν ὑπὲρ ἡμῶν λειφθεῖcα φύcιc. Our hymnist, however, is not necessarily expressing theopaschitism when he asserts that Jesus suffered as ὁ πάντων δεcπότηc. As stated above in our discussion of that ascription, he may only be setting in sharp relief the humiliation of his incarnation over against the majesty of his divine being, a polarity exploited by hymnists and theologians of many doctrinal persuasions.

**27-28 (10) ῥαβδίcαντεc αὐτόν, | cταυρώcαντεc αὐτόν:** ms. ραβδοcεντηc --- cταυρωcαντηc (for ε ⟩ η in the endings of both, cf. Mayser/Schmoll I 1.39-40; Gignac I 244-47; for α ⟩ ε in ραβδ. cf. on πεῖραν [ms. πιρεν] above 24). Aor. forms of the rare ῥαβδίζειν are also attested in LXX Ruth 2.17 (ἐρράβδιcεν) and *P. Ryl.* II 148.20 (ἐράβδιcαν); both mean "thresh (by striking with rod)." There is little evidence for ι ⟩ ο (none cited by Gignac; cf. Psaltes §§ 60, 62, προμοcκρίνιον for πριμιcκρίνιον; ἀρνηcοθεία for ἀρνηcιθεία). The "o" in our word probably arose by analogy with ῥάβδοc. The logical subjects of the nom. participles are the τῶν παρανόμων of the following strophe, in that the latter are considered the real agents of the action; cf. ἐπιγνόντεc δὲ --- φωνὴ ἐγένετο μία ἐκ πάντων, Acts 19.34; Jude 16; Col. 2.2; John 7.38;

cf. KG 2.105-09. For the Trisagion interrupting the syntax between this couplet and the next, see above p. 44.

ῥαβδίζειν means literally "beat with a rod," "cudgel," (cf. Diodor. Sic. *bib.* 19.101.3; 36.2.4; 38/39.8.3; [δερόμενον] ῥάβδοιϲ in P. Berol. inv. 13877.3-4, ed. H. Kortenbeutel, *Aegyptus* 12 [1932] 129ff. with n. p. 132). This Roman punishment was suffered by Paul three times (2Cor. 11.25; cf. Acts 16.22; R. Taubenschlag, *The Law of Greco-Roman Egypt in the light of the Papyri* [Warsaw 1955²] 555f. n. 19) but not, at least in the gospel accounts, by Jesus, who was flogged before the crucifixion (φραγελλοῦν in Mat. 27.26; Mk. 15.15; μαϲτιγοῦν in Jn. 19.1 and the prophecies in Mt. 20.19, Mk. 10.34, Lk. 18.33; cf. Jos. *bel.* 2.308, 5.449).[108] Thus, outside its obvious benefit to the acrostic, the word seems out of place here (the only parallel I can find is Follieri III 391, ῥαβδιζομένου οὐκ ἐφείϲαϲθε). We must, however, take into account that the Greek and Latin terms for various kinds of beatings were notoriously vague and interchangeable (W. Waldstein in *RAC* IX 469-71 s.v. Geißelung; R. Brown, *The Death of the Messiah* I [New York 1994] 851ff.; cf. the discussion of ῥαπίζειν below). ῥαβδίζειν may therefore be intended in a more general sense of "beat" or "thrash."

ῥαβδίζειν, however, may not refer to the Roman scourging before the crucifixion, but to the mockery and beating which Jesus received from the Jews before the Roman trial, which Matthew describes as follows: τότε ἐνέπτυϲαν εἰϲ τὸ πρόϲωπον αὐτοῦ καὶ ἐκολάφιϲαν αὐτόν, οἱ δὲ ἐράπιϲαν (26.67; cf. Mk. 14.65, ῥαπίϲμαϲιν αὐτὸν ἔλαβον); cf. from the mockery of Samson, καὶ ἔπαιζεν ἐνώπιον αὐτῶν, καὶ ἐρράπιζον αὐτόν (LXX Jud. 16.25 B).[109] Like ῥαβδίζειν, ῥαπίζειν in most secular literature means "beat with rods or clubs." Although ῥαπίζειν may also mean slapping, and most modern commentators so interpret it with respect to

---

108On the Roman instruments of beating and their usage determined by class and race, see P. Garnsey, *Social Status and Legal Privilege in the Roman Empire* (Oxford 1970) 136ff.

109Vaticanus' καὶ ἐρράπιζον αὐτόν is not in the Hebrew.

the Jews' treatment of Jesus,[110] our hymn with ῥαβδίζειν understands the tradition at this point as referring to *virgis caedere*, which may indeed be the correct interpretation of Matthew's ῥαπίζειν (see Bauer/Aland s.v.).

**30-31 (11-12) ταῦτα πάντ⟨α⟩ ἔπαθεν | ὑπὸ τῶν παρανόμων:** ms. πάντων, from which I suggest πάντα. However, α ⟩ ω is rare (Gignac I 288; Psaltes §§ 14-16) and we would also have to assume a superfluous ν (Gignac I 112f.). For the sense cf. Ign. *Smyrn.* 2.1, ταῦτα γὰρ πάντα ἔπαθεν δι' ἡμᾶϲ. Another possibility is πάντωϲ (ν ⟩ ϲ; cf. above 23 δεϲπότηϲ, ms. -ην and 14 'Ιωάννηϲ, ms. -ην); cf. the wording in Ath. *gent.* 22.22f. Thompson (Migne PG 25.44d), who argues that the pagan notion that the divine nature possess hands and feet is blasphemy, ἀκολουθεῖ γὰρ αὐτῷ καὶ τὰ ἄλλα τοῦ ϲώματοϲ πάντωϲ πάϲχειν. The reading πάντωϲ is easier paleographically, but πάντα yields better sense, and since α ⟩ ω is paralleled in line 43 (τώ, i.e, τά) I read πάντα here. The rhythm of these lines is also problematic; 30 has 3 stresses, 31 one. The couplet, however, yields 4 stresses, as do the others. Similarly, a seventh-century isosyllabic Easter hymn (*P. Köln* IV 173) displays the principle, that within a given strophe the first colon may have one or two extra syllables if another colon lacks the same number of syllables, so that the strophe maintains the same syllable count as the others (see the metrical text, lines 11, 16 and 17 [p. 66f.] with Römer's comm. p. 70). A similar principle may be at work here with regard to stresses, but I know of no parallels.

The thought resembles that of NT Heb. 12.3, which describes Jesus as τὸν τοιαύτην ὑπομεμενηκότα ὑπὸ τῶν ἁμαρταλῶν --- ἀντιλογίαν.

**33 (12) φωνὴ ἐξ οὐρανοῦ:** a phrase which frequently occurs in NT Revelation in the form of ἤκουϲα φωνὴν ἐκ τοῦ οὐρανοῦ and the like (10.4, 14.2, 18.4; cf. 10.8, 11.2; v.l. in

---

[110]So Schleusner in his *Lexicon in LXX* III s.v. (Glasgow 1822). Schleusner's evidence and discussion does not include Vaticanus' reading of Jud. 16.25 cited above, since that ms. was not edited until 1889-90.

19.5, 21.3). Sometimes its application is obscure (cf. Charles on 10.4 [I p. 262], 18.4 [II p. 97]). For the syntax of this phrase, cf. next n.

**33-37 (12-13) φωνὴ ἐξ οὐρανοῦ, | χαίροντες οἱ ἄγγελο̣ι̣ --- ψάλλοντες καὶ λέγοντες· | ὦ ἀθάνατος κτλ.:** ms. line 34 χυρωντες (αι ⟩ υ Gignac I 275 [one ex., χυρήμωνος for χαιρήμονος], but probably scribal, given the graphic similarity of αι and υ in this and other hands [see scanned images, pp. 32ff., and plate; cf., e.g., our υ here with αι in ηcαιαc and προφηται, dipl. 15; και, dipl. 17; the same confusion caused omission of καί in 45 below]); ms. εαγγελο̣ι̣ (the last two letters might be described as a dot and a downward hook. -οι is the most likely; -ω⟨ν⟩ is also possible [see below]; οι ⟩ ε Gignac I 274f. and cf. below επιcτη for οἱ πιcτοί, 65 [dipl. 29]).

The letters χυροντεcε could also be interpreted as a finite verb and pronoun, yielding two other possibilities: φωνὴ ἐξ οὐρανοῦ, χαίρονταί cε ἄγγελο̣ι̣ ψάλλοντεc καὶ λέγοντεc (for the late middle χαίρομαι, cf. Psaltes § 362, p. 247); or φωναὶ ἐξ οὐρανοῦ χαίρονταί cε ἀγγέλω̣⟨ν⟩ ψάλλοντεc καὶ λέγοντεc (αι ⟩ η Gignac I 248). I prefer, however, the reading given in the lemma, understanding φωνὴ ἐξ οὐρανοῦ as a kind of exclamation (as if a word like ἰδού were present) with the following participial phrases loosely dependent on it. I know of no precise parallels, but we find similar patterns in the language of visionaries; e.g., Rev. 4.1, μετὰ ταῦτα εἶδον, καὶ ἰδοὺ θύρα ---, καὶ ἡ φωνὴ ἡ πρώτη ἣν ἤκουcα ὡc cάλπιγγοc λαλούcηc μετ' ἐμοῦ λέγων, 'Ανάβα ὧδε κτλ. Freedom of participial usage often reigns in the presence of words like φωνή. Cf. also Acts 19.34, ἐπιγνόντεc δὲ ὅτι 'Ιουδαῖόc ἐcτιν, φωνὴ ἐγένετο μία ἐκ πάντων ὡc ἐπὶ ὥραc δύο κραζόντων. Similarly, Jude 16; Rev. 11.15; LXX Gen. 15.1. See in general Mitsakis §§ 305-08, esp. 307; Charles vol. I cxlii § 9; in our document above on lines 27-28). The resulting sequence, χαίροντεc, ψάλλοντεc καὶ λέγοντεc, is a fairly frequent coordination in the *Canticles* of Romanos, with connective only before the last participle (Mitsakis §310). For another acrostic poem with a similar ending, see the anon. Palm Sunday hymn, line 28, in Pitra 476f. (reed. by Römer in *P. Köln* IV p. 71f.): ψάλλοντες καὶ λέγοντες ὡσαννά, υἱὲ Δαυίδ.

The φωνή seems to be the angelic proclamation which greeted Christ at his ascension into heaven.[111] Tradition has the angels applying LXX Ps. 23.7-8 to Jesus on this occasion, with the angels accompanying him in his upward journey saying ἄρετε πύλας, οἱ ἄρχοντες ὑμῶν, καὶ ἐπάρθητε, πύλαι αἰώνιοι, καὶ εἰσελεύσεται ὁ βασιλεὺς τῆς δόξης. The angels in heaven respond, τίς ἐστιν οὗτος ὁ βασιλεὺς τῆς δόξης; to which the first group replies κύριος κραταιὸς καὶ δυνατός, κύριος δυνατὸς ἐν πολέμῳ. Cf. Gr. Naz. *or.* 45.25 (Migne PG 36.657b-c); Gr. Nyss. *ascens.* IX p. 326, 1ff. (Migne PG 46.693a); Ath. *exp. Ps.* 23.7 (Migne PG 27.141c-d). Cf. also Just. *1 apol.* 51.6-7 (Migne PG 6.404b), *dial.* 36.4-6 (Migne PG 6.553c-556a), ibid. 85.1-4 (Migne PG 6.676b-677a). See in general Lampe s.v. ἄγγελος II H2e; J. Michl in *RAC* V 143f. s.v. Engel; P. Beskow, *Rex Gloriae* (Uppsala 1962)103-06; MacCulloch 333f. The original acrostic, prior to its revamping as a Trisagion,[112] may have ended, ὦ (οἱ) ἄρχοντες, ἄρετε πύλας, κτλ. Replacing the Psalm with the ἀθάνατος of the Trisagion would not have appeared out of place. The Egyptian and other eastern traditions viewed both as proclamations of Jesus' victory over death, and in the orientalizing Gallican liturgy the Trisagion and Ps. 23 are closely correlated: *Tunc in adventum sancti evangelii*[113] *claro modolamine denuo psallit clerus "Aius"* (sc. *hymnum trisagium*) *in specie angelorum ante faciem christi ad portas inferni*

[111]This motif is most likely reflected in the NT; cf. I Tim. 3.16, where Christ is said to be ὃς ἐφανερώθη ἐν σαρκί, ἐδικαιώθη ἐν πνεύματι, ὤφθη ἀγγέλοις, κτλ. (J.L. Houlden, *The Pastoral Epistles* [Penguin Books 1976] ad loc., p. 86).

[112]On the general theme of *contaminatio*, see above p. 44f. The second hymn also has the angels singing the ecclesiastic Trisagion. See the general introduction to that section which follows.

[113]For the recitation of the Trisagion before the gospel reading as a commemoration of Jesus' victory over death, cf. above p. 15.

*clamantium "Tollite portas principes vestras et elevamini porte aeternales et introibit dominus virtutum rex gloriae."*[114]

If I have understood this general context correctly, the first hymn concludes with a ring composition effect. The ἀϲτὴρ ἐξ οὐρανοῦ heralds the birth of Christ; the φωνὴ ἐξ οὐρανοῦ greets him as he makes his reentry into the heavenly sphere. And so the worshipper is ushered into the divine throne room where Isaiah saw and heard the seraphim sing the first Trisagion, as the next hymn describes.

## Hymn II, 40-54/15-22

In this section a theological preface on Isaiah's vision of the seraphim (40-45; in general see above pp. 7ff.) precedes the hymn itself (46-54), similar to the panoramic descriptions of celestial worship which precede the angelic Trisagion in liturgies (see *lit. Clem.*, Brightman 18.25ff. and other texts cited above p. 8 with n. 20). This structure underscores what has already been said about the nature of the entire text, that although it is not itself a liturgy for public church service, it sufficiently echoes liturgical style so as to suggest communal rather than private use (see above p. 22). More specifically, parts of the introduction (43-45) and the two couplets of the hymn (47f.; 50f.) are paralleled in eastern baptismal rites (for baptismal/Epiphany motifs in general, see intro. above pp. 22-25). Whereas Isaiah's vision occurred in time (εἶδεν, 42), the angelic adoration, which human worship imitates, is timeless and perpetual (ὑμνοῦϲιν, 43; for a similar use of the present tense, cf. above on 1-2, p. 49). Of the three hymns on our document, this one alone (i.e. the actual hymn, not the preface) has a consistent syllabic as well as accentual pattern (see above pp. 1-5, esp. 4f.).

---

[114]*expositio antiquae liturgiae Gallicanae* § 10 (ed. E.C. Ratcliff, London 1971, p. 7). As this text shows, tradition also applies Ps. 23 to Jesus' "Harrowing of Hell," the dialogue being between Hades and the angels who accompanied Christ in his descent. In general see Beskow, op. cit. 104; *Gospel of Nicodemus* 5 (cited by MacCulloch 164f.) with MacCulloch's n. p. 329f. On the Gallican passage cf. J. Kroll, *Gott und Hölle* (Leipzig 1932) 48; Quasten 60.

The language of the preface is thoroughly Trinitarian (see commentary on the particulars), whereas the subsequent hymn, juxtaposing creation and incarnation themes, is thoroughly Christological. We may note in this connection, that when the Monophysite Severus assumed the see of Antioch in 512, he promoted the Christological Trisagion with the addition so objectionable to orthodox theologians, ὁ cταυρωθεὶc δι' ἡμᾶc. He argued the matter by a distinction: the seraphim indeed address their song to the Trinity, but the church sings its Trisagion to Jesus alone, and thus with the addition retained (cf. *fragmenta* of Severus cited by Migne in his Preface to Jo. D., *trisag.*, PG 95.19-20). Whether or not from Severus' influence, most of the Oriental churches eventually adopted this understanding of the two hymns (see Brock 30f.).

This section of the Michigan Trisagion corresponds to the above-described orientation only to a degree. On the one hand, we do have a prologue which describes the song of the seraphim to the Trinity followed by an ecclesiastic hymn dedicated to Christ. On the other hand, the very hymn which the seraphim sing is the subsequent ecclesiastic hymn with the ascriptions ἰcχυρόc and ἀθάνατοc, but with the threefold ἅγιοc instead of one, more closely resembling the acclamation of Isaiah 6.3. The two traditions are here conflated in a manner which I have found nowhere else; the angelic and ecclesiastic Trisagia coalesce into one hymn in which angels sing that which is really applicable to the human situation (cf. above on 33-37 p. 65 with n. 112, and below on 47f., ὁ κλίναc τοὺc οὐρανοὺc καὶ κατῆλθεν πρὸc ἡμᾶc). We may have here another case of contamination (see above p. 44f.), but on the other hand, this synthesis well illustrates a principal mentioned above in the introduction. The liturgies and popular hymns of Egyptian Christianity sometimes unite the Christological and Trinitarian perspectives in a way that may have been closer to the original situation. Cf. W.E. Crum, *Catalogue of the Coptic Manuscripts in the Collection of the John Rylands Library* (Manchester, 1909) p. 12 no. 34 (van Haelst 990; cf. Schermann 222; Koenen 41) where a Christological Trisagion containing both ὁ

--- ἄνθρωπος γεγονώς and ὁ σταυρωθεὶς δι' ἡμᾶς concludes with a Trinitarian doxology, δόξα καὶ νῦν, ἁγία τριάς (cf. also *lit. Copt. Jac.*, cited above p. 19). The union of the two perspectives also conforms to Patristic interpretations of Is. 6.3 (see below 47-48 on ὁ κλίνας κτλ.).

**40 (15) Ἡσαΐας, ὁ ἐν προφήται⟨ς⟩ μεγαλόφωνος:** The high view of Isaiah is common: *Esaias ... de Christo et ecclesia, ... multo plura quam ceteri prophetavit*, Aug. *de civ. dei* 18.29 with J.F.C. Weldon's commentary ad loc. (II p. 346); cf. idem *conf.* 9. 5.13; *(Esaias) non prophetiam mihi videtur texere, sed Evangelium*, Hier. *ep*. 53.8 (Migne PL 22.547); idem *com. in Is. prol.* (CC 72.1.18ff. [Migne PL 24.18]). With the specific wording of our text cf. ὁ τῶν προφητῶν μεγαλόφωνος κραυγάζει, [Rom.] 29.55 (Pitra 233); Ἡσαΐα θεοκήρυξ, --- τὴν τοῦ θεοῦ φωνήσας μεγαλοφώνως σάρκωσιν, *de Is. proph.* α', Pitra 454; [τὸ]ν μεγαλόφωνον Ἡσαΐαν, Didym. *in Zach.* 3.228; μεγαλοφώνους --- τὰς προφητείας, Didym. *in 2 Reg.* 22.14 (Migne PG 39.1117c). μεγαλόφωνος can have either a positive ("eloquent") or negative ("bombastic") force. Its usage and meaning with regard to Isaiah are probably influenced by the phrase (ἐν) φωνῇ μεγάλῃ, which characterizes the celestial worship of the angels and other beings who surround the throne of God in Revelation (5.2, 12; 6.10; 7.2, 10). He is thus characterized as the one who shares the same uncorrupted vision of God as the angels; with them and like them he worships and proclaims God's true nature "with a mighty voice."

**41-42 (15-16) ὁ ὑπὲρ πᾶσαν τὴν κτίσιν γινώσκων | εἶδεν τὴν ἄκτιστον φύσιν:** The basic thought is Jewish; cf., e.g., Jn. 12.41, ταῦτα εἶπεν Ἡσαΐας, ὅτι εἶδεν τὴν δόξαν αὐτοῦ, καὶ ελάλησεν περὶ αὐτοῦ. The reference here is no doubt to the prophet's vision in Is. 6, where in John's perception it is Jesus who is the δόξα or "dwelling place" (Heb. *shekinah*; cf. Jn. 1.14) of God which Isaiah saw (R. Brown, *The Gospel According to John* [New York, 1987$^2$] ad loc. [I p. 486f.]). The specific phraseology of our hymn, however, reflects a thoroughly Platonizing theological tradition; cf. Philo's praise of Moses as the "more perfect mind" (than the one who knows God through creation), ὅcτιc οὐκ ἀπὸ τῶν γεγονότων τὸ αἴτιον γνωρίζει, ὡc ἂν ἀπὸ cκιᾶc τὸ μένον, ἀλλ' ὑπερκύψαc τὸ γενητὸν ἔμφαcιν ἐναργῆ τοῦ ἀγενήτου λαμβάνει, ὡc ἀπ' αὐτοῦ αὐτὸν καταλαμβάνειν καὶ τὴν cκιὰν αὐτοῦ, ὅπερ ἦν τόν τε λόγον καὶ τόνδε τὸν κόcμον (*leg. al.* 3.100).

**42 (16) τὴν ἄκτιcτον φύcιν:** The phrase is applied to the Trinity (τῆc μὲν ἀκτίcτου φύcεωc τὴν ἁγίαν τριάδα εἶναι διωριcάμεθα, Gr. Nyss. *Eun.* 1.295 [I p. 113.26ff., Migne PG 45. 341c]) or individual members of it, especially Christ (τὸν υἱὸν καὶ τὸ πνεῦμα --- τῆc ἀκτίcτου φύcεωc, Gr. Nyss. *Eun.* 1.220 [I p. 90,22f., Migne PG 45.317a]; ἀκτίcτου φύcεωc, Cωτήρ, καὶ δόξηc ἀϊδίου, [Rom.] Pat. cod. 212, 305.2 [N.B. Tomadakis, Ῥωμανοῦ τοῦ Μελῳδοῦ ὕμνοι, Athens 1952-61, II p. cμβ']), esp. when his uncreated nature is contrasted with his created, i.e., his body (Procl. CP *or.* 2.2, Migne PG 65.693B; Apoll. *corp. et div.* 4f. p. 187.2ff. Lietzmann [Migne PL 8.873c]); see Lampe s.v. ἄκτιcτοc 4C.

**43-45 (17-18) καὶ τὰ cεραφὶμ ὑμνοῦcιν | τρι{α}cὶν ἁγιαcμοῖc | εἰc μίαν οὐcίαν ⟨καὶ⟩ κυριότητα:** ms. τ ω cεραφιν; α ⟩ ω is rare and almost always before a nasal (Gignac I 288; Psaltes §§ 14-16; cf. above on 30 ms. παντων, i.e., πάντα). Ms. τρι{α}cιν αγιαcμοc ({α} possibly because of the graphic similarity between -αcι and the following αγι-; for -οc as dat. plu., see Mayser/Schmoll I 1.88; cf. Gignac I 199ff.). The graphic similarity of κυ- and καί caused the omission of the latter (cf. above 33-36 on χύροντεc). However, another possibility is εἰc μίαν οὐcίαν, ⟨μίαν⟩ κυριότητα, which would resemble the phraseology of the Hippolytan baptismal liturgy (reconstructed from the Coptic and other versions by Lietzmann, *Symbolst.* 76/264), τριάδα ὁμοούcιον, μίαν θεότητα, μίαν κυριότητα. For the reading of the lemma, cf. Jo. D. *trisag.* 12.15ff. cited below. Rhythmically either reading works. The εἰc μίαν οὐcίαν, ⟨μίαν⟩ κυριότητα option has four stresses and could thus function as a clausula for the introduction (on clausulae cf. above p. 2).

The phraseology liturgically mirrors orthodox dogma, that the Trinity was of three distinct persons and of one substance or being. Cf. οὕτω μὲν οὖν τὰ ἅγια τῶν ἁγίων[115] --- δοξάζεται τριϲὶν ἁγιαϲμοῖϲ εἰϲ μίαν ϲυνιοῦϲι κυριότητα καὶ θεότητα (Gr. Naz. *in theophan.* [or. 38], Migne PG 36.2.320; so also *in sanct. pasch.* [orat. 45], Migne PG 36.2.628f., and cited by Jo. D. *expos. fid.* 54.33ff. and *trisag.* 22.8ff.); τριϲὶν ἁγιαϲμοῖϲ ὑμνεῖν τε καὶ ἁγιάζειν τὴν τριϲυπόϲτατον μίαν θεότητα διδαχθήϲεται τῶν ἀνθρώπων ἡ φύϲιϲ (Max. *myst.* 19, Migne PG 91.696c); ἑνὰϲ οὐϲίαϲ καὶ μία κυριότηϲ τῷ τριϲϲῷ ἁγιαϲμῷ δηλουμένη ἡ τριττὺϲ τῶν ὑποϲτάϲεών ἐϲτιν. ὥϲτε τὸ "ἅγιοϲ ἅγιοϲ ἅγιοϲ" τῶν τριῶν ὑποϲτάϲεών ἐϲτιν ἐνδεικτικόν, οὐ μιᾶϲ, τὸ δὲ "κύριοϲ Caβαωθ" τῆϲ μιᾶϲ θεότητοϲ καὶ κυριότητοϲ ἐμφαντικόν (Jo. D. *trisag.* 12.15ff. [Migne PG 95.45b]). In the preceding material τριϲὶν ἁγιαϲμοῖϲ (cf. in the last citation τῷ τριϲϲῷ ἁγιαϲμῷ) designates the angelic Trisagion; for its use for the ecclesiastic hymn, cf. Sophr. H. *or.* 2.5, Migne PG 87.3.3224a, ϲταυρὸν ἀϲεβῶϲ τοῖϲ τριϲὶν ἁγιαϲμοῖϲ ἐπιφέρουϲιν (cf. also 2.3, ibid. 3220c). The specific formula of Trinitarian unity μία οὐϲία καὶ κυριότηϲ *vel sim.* was favored by Jo. D.; e.g, *trisag.* 8.18f. (Migne PG 95.41b), τὸν ἐμφαντικὸν τῶν τριῶν ὑποϲτάϲεων καὶ τῆϲ μιᾶϲ οὐϲίαϲ καὶ κυριότητοϲ αἶνον. Cf. ibid. 3.29 (PG 95.28d), 10.14f. (PG 95.44b), 24.9 (PG 95.56a). For the interpretation in many Oriental churches of the angelic Trisagion as Trinitarian and the ecclesiastic hymn as Christological, see the introduction to this section.

**46 (18), 49 (19), 52 (21) ἅγιοϲ ἅγιοϲ ἅγιοϲ:** applied to Jesus alone; cf. ἅγιοϲ ἅγιοϲ ἅγιοϲ εἶ ὁ καθήμενοϲ ἐν δεξιᾷ τοῦ πατρόϲ, Maas *FK* I 5.20f.; similarly τριϲάγιε ϲωτὴρ ἡμῶν, ibid. 6.1. This especially offended the rigid Trinitarian position of John of Damascus. He says that, after the congregation recites the Trisagion, ὁ ἱερεὺϲ ὥϲπερ ἑρμηνεύων τὸν ὕμνον φηϲίν· "ἅγιοϲ εἶ, βαϲιλεῦ τῶν αἰώνων, --- ἅγιοϲ καὶ ὁ μονογενήϲ ϲου υἱόϲ, --- ἅγιον δὲ καὶ τὸ πνεῦμά ϲου --- ." διὰ τί μὴ λέγει ἅγιοϲ, ἅγιοϲ, ἅγιοϲ εἶ, βαϲιλεῦ τῶν αἰώνων, ἅγιοϲ, ἅγιοϲ, ἅγιοϲ, ὁ μονογενήϲ ϲου

[115] τὰ ἅγια τῶν ἁγίων is here equivalent to ἡ ἁγία τριάϲ.

υἱός, ἀλλὰ κοινῶς μὲν τῇ τριάδι "ἅγιος, ἅγιος, ἅγιος," καθ' ἑκάστην δὲ τῶν ὑποστάσεων ἅπαξ τὸ ἅγιος; (*trisag.* 27.3-9 [Migne, PG 95.57c-d]). On this structure see also the introduction to this section, above pp. 67f.

**47-48 (18-19) ὁ κλίνας τοὺς οὐρανοὺς | καὶ κατῆλθεν πρὸς ἡμᾶς**: adapted from LXX Ps. 17.10, καὶ ἔκλινεν οὐρανὸν καὶ κατέβη (many mss. read οὐρανούς); similarly, κύριε κλῖνον οὐρανούς σου καὶ κατάβηθι, ibid. 143.5; cf. also Jesus' statement, καταβέβηκα ἀπὸ τοῦ οὐρανοῦ, Jn. 6.38. I can find no variants for κατέρχεσθαι instead of καταβαίνειν in these Biblical texts. The former occurs in our document possibly under the influence of the creeds, which use this verb for Jesus' descent (Lietzmann, *Symbolst.* 15f./203f. and below on κατῆλθεν κτλ., 68-70).

We could normalize the grammar by understanding {καὶ}, which perhaps intruded from the Psalm, or by retaining καί and reading κατελθών (ε ⟩ η also occurs elsewhere in this text and is generally common [see appendix; Gignac I 244ff.]; ω ⟩ ε, does not occur and is generally rare [Gignac I 292]). The cola, however, make sense as they stand. Various kinds of coordination of participles with finite verbs occur commonly enough in classical (KG 2.100f.; Schwyzer II 406) and late Greek (H. Frisk, *Glotta* 17 [1929] 56ff.). The particular form, however, that it takes in our text, i.e., an articular participle coordinated with a finite verb, is found mainly in Greek with a Semitic coloring, such as the LXX (e.g., ὁ θεὸς ὁ περιζωννύων με δύναμιν καὶ ἔθετο ἄμωμον τὴν ὁδόν μου, ὁ καταρτιζόμενος --- καὶ ἔθου --- κτλ., Ps. 17.33ff.), NT (e.g., τῷ ἀγαπῶντι ἡμᾶς καὶ λύσαντι ἡμᾶς ἐκ τῶν ἁμαρτιῶν ἡμῶν ἐν τῷ αἵματι αὐτοῦ, καὶ ἐποίησεν ἡμᾶς βασιλείαν κτλ., Rev. 1.5f.; cf. Col. 1.26, BDR § 468.3) and elsewhere in hymns (e.g., ἕως γεννήσῃς [scil. Ζαχαρίας] φωνὴν τὴν λύσασαν τῆς ἀφωνίας σὲ τὸν γεννήσαντα καὶ ὑποδείξει Χριστὸν ἀνθώποις (inst. of τὴν ὑποδείξασαν κτλ.), Rom. *cant. dub.* 61ις' [bis] 5-7, p. 16; cf. Mitsakis § 308). For this idiom and its Semitic background, see J.H. Moulton, W. F. Howard, *A Grammar of New Testament Greek* II (Edinburgh 1929) 428f.; Norden, *Ag. Th.* 387. Norden's explanation of Rev. 1.5f., that the relative pronoun was omitted through carelessness, does not explain the phenomenon in general. It seems more precise to say that in the Semitic style of piling up attributes (see Norden, ibid. 201f.) the structural distinction between participles and finite verbs is sometimes lost and the mind can jump quickly from one to the other. For an extreme example, cf. Rev. 1.4, ὁ ὢν καὶ ὁ ἦν καὶ ὁ ἐρχόμενοϲ (with Charles ad loc., I p. 10; cf. also the Trisagion of Rev. 4.8, above p. 11).[116]

In this first couplet of hymn 2 we immediately move from the Trinitarian focus of the preface on the vision of Isaiah (see intro. to this section above) to a Christological statement of the incarnation. We find a similar flow of thought in a Patristic exegesis of Is. 6.3, which maintains that, although with ἅγιοϲ ἅγιοϲ ἅγιοϲ κύριοϲ Cαβαωθ the angels designate the Trinity in both its plurality and unity, in the phrase πλήρηϲ ὁ οὐρανὸϲ καὶ ἡ γῆ τῆϲ δόξηϲ ϲου[117] they acknowledge the manifestation of heavenly glory on earth, and so establish a transition between the worship of the Trinity and that of the incarnate Christ: τὸ μὲν γὰρ ἅγιοϲ ἅγιοϲ ἅγιοϲ τῆϲ τριάδοϲ δηλωτικόν, τὸ δὲ κύριοϲ Cαβαὼθ τῆϲ φύϲεωϲ τῆϲ μιᾶϲ ϲημαντικόν. ὑμνεῖ δὲ τὰ ϲεραφὶμ τὴν ἀΐδιον φύϲιν ὡϲ μὴ μόνον τὸν οὐρανὸν ἀλλὰ καὶ τὴν γῆν ἅπαϲαν τῆϲ δόξηϲ ἐμπλήϲαϲαν. τοῦτο δὲ ἡ τ(οῦ) θεοῦ καὶ (ϲωτῆρο)ϲ ἡμῶν πεποίηκεν ἐνανθρώπηϲιϲ (Thdt. *Is.* 3.70-75 [SC 276 pp. 260f.]). Similarly, Eusebius says of the seraphim, οὐ μίαν δὲ ἠφίεϲαν φωνὴν οὐδὲ ἀθρόωϲ ὁμοῦ τὰ πάντα κοινῇ ταύτην ἀνέπεμπον, ἀλλ' ἕτεροϲ πρὸϲ τὸν ἕτερον ἐμφαίνων τὸ ἑαυτοῦ θαῦμα καὶ τὴν ἔκπληξιν τὴν ἐπὶ τῷ θεωρουμένῳ. μάλιϲτα δὲ ὑπερεξέπληττεν αὐτὰ ἡ ἀπὸ τῶν ὑψηλοτέρων ἐπὶ τὰ ταπεινὰ κατάβαϲιϲ τῆϲ δόξηϲ τοῦ θεοῦ λόγου. τοῦτο γοῦν αὐτοῖϲ παρεῖχεν θαῦμα μέγιϲτον, ὅτι μὴ μόνοϲ ὁ οὐρανὸϲ πλήρηϲ ἐτύγχανεν τῆϲ δό-

[116]Cf. also A. Dihle's discussion of various passages in Trypanis, "Three Hymns," # 1 (*BZ* 69 [1976] 2). Here we are dealing with a somewhat different phenomenon of a protasis consisting of article + participle followed by an imperitival apodosis degenerating into a protasis with article + finite verb. Our hymn has several examples of the correct construction (see above p. 19 n. 54).

[117]i.e., the liturgical adaptation of the actual Septuagint phrase πλήρηϲ πᾶϲα ἡ γῆ τῆϲ δόξηϲ αὐτοῦ. See above p. 7.

ξηϲ αὐτοῦ, ἀλλ' ἤδη καὶ μέχρι γῆϲ ἔφθανεν, ὡϲ πληροῦϲθαι τὴν ϲύμπαϲαν γῆν τῆϲ αὐτοῦ δόξηϲ (*comm. in Is.*, ed. J. Ziegler, *Eusbius Werke*, 9. Band [GCS, Berlin 1975] p. 39,25ff.); cf. Brock 27 with nn. 11-12.

The Biblical phrase "who bowed the heavens and came down" occurs frequently in Coptic liturgies in general (see above p. 25 n. 70), including three times in the baptismal office: once before the prayer of consecration of the oil, and twice introducing the second of the three "prayers of absolution" (Burmester, *Coptic Church* 115, 121, 126; cf. idem, "Baptismal Ritual," 51, 56 [cf. 74], 82; Woolley, *Coptic Offices* 11, 39, 55. On the prayers of absolution see Burmester, *Coptic Church* 330). The idea of *katabasis* in fact closely links the redemptive significance of Jesus' incarnation with that of his baptism: he descended into human flesh to save it; he descended into the waters of chaos to sanctify them (see on lines 68-70 below). In general, for the Epiphanal or baptismal themes in our text, see intro. above, pp. 22-25.

**50-51 (20) ὁ πλάϲαϲ τὸν ἄνθρωπον | κατ' ἰδίαν εἰκόνα:** ms. καθ' ιδιαν. We may describe the *koine* aspirated spelling ἵδιοϲ on the basis of an original development, which accounts for the great majority of the instances, and then on the basis of two fairly minor developments from that point of origin. It first emerged in connection with the idiom καθ' ἰδίαν, "privately," which occurs frequently in inscriptions (81 examples among those catalogued on PHI CD ROM 7; cf. Threatte, *Grammar* I 501) and is usually explained by analogy with phrases such as καθ' ἑαυτόν and καθ' ἕκαϲτον (BDR § 15.2; Schwyzer I 305 with n. 4). It is also possible, however, that the aspirate arose from its direct association with ἕκαϲτοϲ in phrases such as καὶ κοινῇ καὶ καθ' ἰδίαν ἑκάϲτῳ, καθ' ἰδίαν ἑκάϲτῳ καὶ κατὰ κοινὸν, *vel sim.* (*OGIS* I 233.45 [*Inscr. Mag.* 61; reign of Antiochos III]; IG IX 1 278, 5 [II BC ]; IG XII 5 860. 3 [I BC]; cf. D.Chr. *or.* 32.12). In time it extended more generally to καθ' ἰδίαν _______ (i.e., to instances where the phrase is not an independent idiom, but where ἰδίαν modifies another noun) and to the form ἰδίαν without καθ' (e.g., ᾐχμαλωτεύθηϲαν εἰϲ γῆν οὐχ ἰδίαν, LXX Judith 5.18 SA). To this second stage our καθ' ἰδίαν εἰκόνα belongs, as does the one other instance of ἴδιος which I can find in documentary papyri, καθ' ἰδίαν μῆνιν,[118] *P. Ryl.* II 67.3 (late II BC). In a third phase of evolution, the aspirate spreads to other forms of ἴδιος besides ἰδίαν and also to compounds (e.g., οὐχ ἰδίων, Hipp. *haer.* VI 27.1 P [Marcovich p. 234]; οὐχ ἴδια κτήματα, Ph. *spec. leg.* 4.72 $S^1$; καθ' ἰδιόγραφον, *PSI* XII 1235.24 [86-89 AD]; *P. Oxy.* I 70.6f. [III AD]). Although we may phonetically explain the development of the phenomenon in this way, by the period of our text the distinction between καθ' ἰδίαν/κατ' ἰδίαν, οὐχ ἴδιος/οὐκ ἴδιος *vel sim.* may have been largely orthographic, since the initial aspirate had ceased to be pronounced in the speech of many writers (Gignac I 133).

For the various texts above cited on ἴδιος, see BDR § 14.2; J.H. Moulton, W.F. Howard, *Grammar of New Testament Greek* II 98; H.J. Thackeray, *A Grammar of the Old Testament in Greek* (vol. I; Cambridge 1909) 126; Mayser-Schmoll $I^2$ 1.175; Gignac I 135f.; G. Crönert, *Memoria Graeca Herculanensis* (Leipzig 1903; repr. Hildesheim 1963) 148f. n. 2. In general for "false" or analogous aspiration, see Crönert, op. cit. 146-55; Schwyzer loc. cit.

The cola seem to represent a conflation of the two Genesis accounts of the creation of man: κατ' εἰκόνα θεοῦ ἐποίησεν αὐτόν (1.27); καὶ ἔπλασεν ὁ θεὸς τὸν ἄνθρωπον χοῦν ἀπὸ τῆς γῆς (2.7). Cf. ὁ πλάσας κατ' εἰκόνα ἰδίαν, Follieri III 137; ἄνθρωπον --- ἔπλασεν τῆς ἑαυτοῦ εἰκόνος χαρακτῆρα, 1Clem. 33.4. The attribution, "who made man in his own image," is usually ascribed to God the Father rather than, as here, to Christ (the εἴκων of the Father being Christ, Col. 1.15; cf. 2Cor. 4.4). Our hymnist, however, probably has in mind the basic tenet of logos theology, that whereas the Father plans and prescribes the creation of the kosmos and man in his image, Jesus is the actual *fabricator*; cf.

[118] The reading seems sound, but μῆνις in a documentary text causes pause. I find it nowhere else in documentary papyri (of the those catalogued in PHI CD 7 and lexica) nor in the NT, and Lampe does not list it for Patristic texts. It does occur a few times in the LXX and other Greek versions of the OT, and once in Hermas (*past.* 34.4). There are also some scattered uses of it in late Greek philosophic texts (e.g., *SVF* III 397) and other prose (Jos. *ant.* 9.104).

above on 4-5 γεννᾶται --- διὰ λόγου θεοῦ pp. 50f., and Epiph. *pan.* II (GCS 31) 195.15ff.: ἀλλὰ δῆλον ὅτι αὐτός ἐϲτιν ὁ δημιουργὸϲ τοῦ ἀνθρώπου καὶ τοῦ κόϲμου, ᾧ εἶπεν ὁ πατήρ, "ποιήϲωμεν ἄνθρωπον κατ' εἰκόνα ἡμετέραν καὶ καθ' ὁμοίωϲιν"· ἀπὸ δὲ τοῦ ἑνὸϲ ἔργου ϲυϲταθήϲεται φανερὸϲ γινόμενοϲ ὁ τεχνίτηϲ ὅτι οὗτόϲ ἐϲτιν ὁ τότε τὸν ἄνθρωπον πεποιηκώϲ, πλάϲαϲ τε ἐκ γῆϲ τὸ τοῦ 'Αδὰμ ϲῶμα καὶ ποιήϲαϲ αὐτὸ εἰϲ ψυχὴν ζῶϲαν. διὸ καὶ ἐμαρτύρηϲεν ὁ ἅγιοϲ 'Ιωάννηϲ "--- πάντα δι' αὐτοῦ ἐγένετο ---." εἰ δὲ ⟨πάντα⟩ ἐν αὐτῷ ἐγένετο καὶ ὑπ' αὐτοῦ ἐγένετο, αὐτὸϲ τότε τὸν 'Αδὰμ ἔπλαϲε καὶ αὐτὸϲ πάλιν τὸ ϲῶμα ἀπὸ Μαρίαϲ τῆϲ παρθένου εἰϲ ἑαυτὸν ἀνεπλάϲατο.

In conclusion, Epiphanius' words provide us grist for investigating the issue of what relation there is, if any, between the two couplets of the second hymn, ὁ κλίναϲ τοὺϲ οὐρανοὺϲ καὶ κατῆλθεν πρὸϲ ἡμᾶϲ and ὁ πλάϲαϲ τὸν ἄνθρωπον κατ' ἰδίαν εἰκόνα, and what theological influences may be at work in uniting these ascriptions. In general, this juxtaposing of Christ as creator/pantocrator and Christ as incarnate redeemer suits well the blending of these themes which characterizes the entire document (see general intro. above pp. 21f.). More specifically with regard to the creation of man in God's image, the last sentence of the above Epiphanius quote maintains that the same logos who formed the body of Adam came down and formed a body for himself in the womb of the virgin (again, see above on 4-5 γεννᾶται --- διὰ λόγου θεοῦ p. 52). In addition, Christian thought and liturgy establishes a close correlation of the notions of incarnation and divine image, in that Jesus through his *katabasis* restores the εἰκών from the corruption of sin. Cf., e.g., from the Service of the Catechumenate connected with the Greek baptismal liturgy: ὁ ὤν, δέϲποτα κύριε, ὁ ποιήϲαϲ τὸν ἄνθρωπον κατ' εἰκόνα τὴν καὶ ὁμοίωϲιν --- εἶτα ἐκπεϲόντα διὰ τῆϲ ἁμαρτίαϲ μὴ παριδών, ἀλλ' οἰκονομήϲαϲ διὰ τῆϲ ἐνανθρωπήϲεωϲ τοῦ Χριϲτοῦ ϲου τὴν ϲωτηρίαν τοῦ κόϲμου Goar, *Euchologion* 276f.; Burmester, "Baptismal Rite" 30); cf. Gregory's description of Christ as ὁ ἐνανθρωπήϲαϲ δι' ἡμᾶϲ καὶ πτωχεύϲαϲ θεόϲ, ἵνα --- ἀναϲώϲηται τὴν εἰκόνα καὶ ἀναπλάϲῃ τὸν ἄνθρωπον (Gr. Naz. *or.* 7.23, Migne PG 35.785c).

The two ideas also figure prominently in the Coptic baptismal liturgy, where, in preparation for the neophyte's entering the font, the priest prays: "... that they who are about to be baptized therein, may put off the old man which is corruptible according to the deceitful lusts, and put on the new man which shall renew again according to the image of him that created him" (trans. Burmester "Baptismal Rite" 41, 43; cf. the parallel passage in the Greek liturgy ibid. 40, 42; Goar, *Euchologion* 289). Then immediately follows the recitation of the Lord's prayer and the "three prayers of absolution," the second of which begins with our specific phrase, "he bowed the heavens and came down" (Burmester, "Baptismal Rite" 74; cf. idem, *Coptic Church* 121; on the prayers of absolution see above p. 73 on 47-48). In general, see Lampe s.v. εἰκών C; H. Crouzel in *EEC* I 406 s.v. Image; H. Merki in *RAC* IV 464ff., esp. 471-75 s.v. Ebenbildlichkeit. For baptismal motifs in this document, see above pp. 22-25.

## Hymn III and Concluding Sextuplet, 55-70/22-31

This section consists of the Trisagion expanded by quotes from the LXX on creation themes (55-64). The omission of the expected ἐλέηϲον ἡμᾶϲ after the ἀθάνατοϲ colon (63) is probably intended to provide a smooth transition into the final sextuplet, which concludes the three-hymn cycle with a call to worship the incarnate Christ (65-70). By way of a credal statement the sextuplet reiterates in ring composition fashion the themes of Jesus' birth and baptism which began the first hymn.

The Biblical quotes of hymn 3 present Jesus as the supreme *pantocrator* by way of a polarity: Jesus is the Lord of the heights, who stretches out the heavens and forms the constellations (56, 59-60); he is also Lord of the depths, who treads on the chaotic waters as land and subdues them by setting their boundary beyond which they cannot go (57, 64; the configuration of the ascriptions is thus ABAB). Various eastern liturgies of baptism exploit similar polarities, as, e.g., in the Orthodox rite: ϲὺ ἐξετείναϲ τὸν οὐρανὸν ὡϲεὶ δέρριν· ϲὺ ἐϲτερέωϲαϲ τὴν γῆν ἐπὶ τῶν ὑδάτων· ϲὺ περιετείχιϲαϲ τὴν θάλαϲϲαν ψάμμῳ (Goar, *Euchologion* 352-53; Burmester, (Baptismal Rite" 38). Cf. the parallel passage in the Coptic rite: "Who didst create heaven and earth ... Who didst gather the waters into one gathering place; who didst bind the sea and confine the depths and didst seal them with thy glorious and fearful name" (Burmester, "Baptismal Rite" 37f.). This cosmogonic supremacy, especially over the waters of chaos, is typologically correlated with Jesus' power by his incarnation and baptism to sanctify the baptismal waters for the salvation of humans. This gracious condescension is the focus of the final sextuplet.

In my edited text of the third hymn (see above pp. 40, 42) I have set the LXX material into cola, in accordance with the accentual structure of the other parts of the text. That these passages are Greek translations of Biblical Hebrew poetry helps to justify this approach (see above pp. 2ff.), but I also concede that they may not have been intended to be read rhythmically.

With regard to the LXX material I compare in the following table the version of the standard Göttingen editions with that of the Michigan document.

| ***P. Mich.* 799.57-58, 59-62** | **LXX Job 9.8-10** |
| --- | --- |
| 56 ὁ ἐκτείναϲ τὸν οὐρανὸν ἐν τῇ{ϲ} ϲυνέϲει αὐτοῦ· | ὁ τανύϲαϲ τὸν οὐρανὸν μόνοϲ |
| 57 ὁ περιπατῶν ὡϲ ἐπὶ ξηρᾶϲ ἐπὶ θαλ{λ}άϲ⟨ϲ⟩ηϲ· | καὶ περιπατῶν ὡϲ ἐπ' ἐδάφουϲ ἐπὶ θαλάϲϲηϲ· |
| 59 ὁ ποιῶν τὴν Πλειά{ν}δα καὶ Ἕϲπερον | ὁ ποιῶν Πλειάδα καὶ Ἕϲπερον |
| 60 καὶ Ἀρκτοῦρον καὶ ταμιεῖα νότου· | καὶ Ἀρκτοῦρον καὶ ταμιεῖα νότου· |
| 61 ὁ ποιῶν τὰ μεγάλα καὶ ἀνεξιχνίαϲτα, | ὁ ποιῶν μεγάλα καὶ ἀνεξιχνίαϲτα, |
| 62 ἔνδοξά τε καὶ ⟨ἐ⟩ξαίϲια, ὧν οὐκ ἔϲτιν ἀριθμόϲ. | ἔνδοξά τε καὶ ἐξαίϲια, ὧν οὐκ ἔϲτιν ἀριθμόϲ. |

| ***P. Mich.* 799.63-64** | **LXX Jer. 5.22** |
|---|---|
| 64 ὁ τιθεὶς ⟨ἄμ⟩μον ὅριον ἐν τῇ θαλ{λ}άς⟨ς⟩η{ς} | τὸν τάξαντα ἄμμον ὅριον τῇ θαλλάςςῃ |

**56 (22)** ms. ςενεςει (I thank D. Wilson for the reading ςυνέςει; υ ⟩ ε Gignac I 273f.). Some LXX mss. read ἐκτείνων instead of τανύςας, and cf. ἐγὼ κύριος --- ἐξέτεινα τὸν οὐρανὸν μόνος, Is. 44.24; I find no LXX evidence for ἐν τῇ ςυνέςει αὐτοῦ replacing μόνος. For the thought, cf. 1Clem. 33.3: τῷ γὰρ παμμεγεθεςτάτῳ αὐτοῦ κράτει οὐρανοὺς ἐςτήριςεν καὶ τῇ ἀκαταλήπτῳ αὐτοῦ ςυνέςει διεκόςμηςεν αὐτούς. For similar themes in baptismal liturgies, see intro. to this section.

**57 (23-24)** ms. περιπατωρ (ν ⟩ ρ Gignac I 109); θαλλάςης (see also 64; λ ⟩ λλ Gignac I 155f.; ςς ⟩ ς Gignac I 158f.). No ms. evidence for ξηρᾶς in LXX Job 9.8, but cf. in the "Song of Moses," οἱ δὲ υἱοὶ Ιςραηλ ἐπορεύθηςαν διὰ ξηρᾶς ἐν μέςῳ τῆς θαλάςςης, Ex. 15.19; also Ps. 65.6 (I owe these references to L. MacCoull). The topos of God treading upon the chaotic waters as on dry land both refers typologically to the well-known miracle of Jesus walking on the Sea of Galilee (Mk. 6.45-52 and par.) and adumbrates the baptismal theme explicitly stated in line 70 (on which see below on 68-70 and the intro. to this section above).

**59 (24-25)** ms. πυουν (οι ⟩ υ Gignac I 197f.; ω ⟩ ου Gignac I 209f.); Πλιαντα (development of ν, Gignac I 118 and to his refs. add ἐξίςτα{ν}το, *P. Köln* IV 173.10 (diplomatic, 8 edited; but possibly *constructio ad sensum* [see Römer ad loc.]) and Threatte, *Grammar* I 488f.; δ ⟩ τ Gignac I 80); εςπειρεν (ε ⟩ ει Gignac I 256f.; ο ⟩ ε ibid. 289f.).

**60 (25)** ταμιεῖα ⟩ ταμῖα by contraction (Gignac I 295ff. esp. 296).

**61 (26)** ms. ποιουν (see on 58); ανεξεχνιαςτα, ι ⟩ ε Gignac I 251ff.; -εχ-, however, may be by phonetic assimilation with the previous -εξ-. Our scribe is especially prone to mistakes in words of several syllables; cf. ραβδοcεντηc, above 27 and ανεκτιεκετο, below 67.

**62 (26-27)** ms. ενδοξοι (α ⟩ οι must be scribal since there is no phonological basis for it during any period); ms. δε (initial τ ⟩ δ Gignac I 80f.). Ms. αριθμου for ἀριθμόc probably by contamination with the preceding ὧν.

Two factors may account for the vowel loss in ⟨ἐ⟩ξαίcια: the preceding καί (αι being the same sound as ε) and a growing tendency in Byzantine Greek (firmly established in the modern language) to dispense with unaccented initial vowels (particularly *e*); cf. R. Browning, *Medieval and Modern Greek* (Cambridge 1983²) 57f.; Mandilaras, op. cit. (above p. 57) § 236 with n. 1; A. Mirambel, *Grammaire du grec moderne* (Paris 1949) 27; Jannaris, op cit. (above p. 10 n. 24), §§ 134-35; A. Thumb, *A Handbook of the Modern Greek Language* (trans. S. Angus; Chicago 1964) §12a. Cf. also in the Petra papyri inv. 10, where one always has ξώcτρα instead of the expected ἐξώcτρα (e.g., line 11; corresp. L. Koenen, 2/1/99).

**64 (28)** Basil. Mag.'s quote of this verse, similar to the Michigan document, reads τιθέντα instead of τάξαντα of the Göttingen text. ⟨ἄμ⟩μον by partial haplography (Gignac I 312f.). On θαλ{λ}άc⟨c⟩η{c} see above 56 (superfluous final c Gignac I 125f.). For ἐν + dat. = simple dat. see BDR § 220.1; Mitsakis § 186. For the thought and parallels, see the intro. to this section.

## Concluding Sextuplet

**65 (28-29) δεῦτε οὖν πάντεc οἱ πιcτοί:** ms. ε πιcτη (οι ⟩ ε Gignac I 274f.; οι ⟩ η ibid. 265ff.; neither is frequent in the Roman/Byzantine periods,[119] and Gignac cites Coptic influence for both; on the former see also above 33-36 ε αγγελοι; on the latter see Römer p. 80). Cf. δεῦτε (πάντεc πιcτοὶ) προcκυνήcωμεν τὸν cωτῆρα Χριcτόν, which occurs in songs of the ἀπόδειπνον

[119]Even less so in the Ptolemaic period; cf. Teodorsson 141 (§§ 76, 77) with n. pp. 228f.

(Maas, *FK* I 2.1, 3.1f.) and other hymns (ibid. 7.1). For *P. Mich.* 799 and the ἀπόδειπνον tradition, see above pp. 25ff. Cf. also δεῦτε μετὰ πίcτεω[c, MPER N.S. XVII 22.27. For the coupling of δεῦτε and ἴδετε (see next line) cf. Mat. 28.6, Jn. 4.29; in pagan lit., cf. δεῦρ' ἴδε, Theok. 27.46 with Gow's n. ad loc.

**66-67 (29-30) ἴδετε καὶ θαυμάcατε καὶ ⟨ ⟩ | τὴν ἀνεκδιήγητο⟨ν⟩ ἑαυτοῦ φιλανθρωπία⟨ν⟩:** ms. ανεκτιεκετο (medial δ ⟩ τ Gignac I 81; omission of final ν ibid. 111f.). With regard to -εκε- for -ηγη-, intervocalic γ ⟩ κ is common enough (Gignac I 79), as is η ⟩ ε. The latter, however, tends to occur in specific phonetic conditions (ibid. I 242-44), none of which are met here, and thus our -εκε- may be due to graphic/phonetic assimilation to the previous -εκ-; cf. 61/26 above on ανεξ**εχ**νιαcτα for ἀνεξιχνίαcτα, where -εχ- may be influenced by the preceding -εξ-. For the reflexive ἑαυτοῦ used for the personal pronoun cf. Mayser II 2.70f., Gignac II 170f.

For the general wording and thought, cf. ἴδετε, λαοί, καὶ θαυμάcατε· (Follieri II 164); ἴδετε ποταπὴν ἀγάπην δέδωκεν ἡμῖν ὁ πατήρ, ἵνα τέκνα θεοῦ κληθῶμεν (1Jn. 3.1). The missing verb in line 66 could be, e.g., ἐπιβλέψατε; cf. LXX Hab. 1.5, ἴδετε, οἱ καταφρονηταί, καὶ ἐπιβλέψατε καὶ θαυμάcατε θαυμάcια καὶ ἀφανίcθητε, διότι ἔργον ἐγὼ ἐργάζομαι ἐν ταῖc ἡμέραιc ὑμῶν κτλ., which is quoted in Acts 13.41 and in Patristic lit. (Chrys. *hom. in Ac.* 29.1, Migne PG 60.215.28ff.; Thdt. *Os.–Mal.*, Migne PG 81.1812c-d; et al.). We could, however, also read ἴδετε καὶ θαυμάcατε {καὶ} τὴν κτλ. It is difficult to say which reading works better rhythmically. The latter would yield a consistent pattern for the entire sextuplet, with 3 lines of 3 stresses alternating with 3 of 2. The version of the lemma gives us 3 stresses each for lines 65-67, which form a sense unit (see edited text above p. 42).

In both secular and sacred contexts φιλανθρωπία most frequently signifies the kindness of a benefactor (see Spicq, op. cit. [above p. 60] III 440ff.). As such it sometimes implies a redemptive intervention or condescension, of which, in a Christian context, the incarnation is the example par excellence; thus the following κατῆλθεν πρὸc ἡμᾶc. Cf. ὁ ἐν μεcονυκτίῳ, φιλάνθρωπε, ἐν cπηλαίῳ τεχθῆναι ηὐδόκηcαc, καὶ ἐμὲ --- [ἐν] ἀνομίαιc τεχθέντα ἐλέηcον, Trypanis, "Three Hymns," 1.19ff.; similarly Philo says of Yahweh, *de cher.* 99, εἰ γὰρ βαcιλεῖc ὑποδέχεcθαι μέλλοντεc λαμπροτέραc καταcχευάζομεν τὰc ἰδίαc οἰκίαc --- , τῷ βαcιλέων βαcιλεῖ καὶ τῶν cυμπάντων ἡγεμόνι θεῷ δι' ἡμερότητα καὶ φιλανθρωπίαν ἀξιώcαντι τὸ γενητὸν ἐπιcκέψεωc καὶ ἀπ' οὐρανοῦ περάτων μέχρι γῆc ἐcχάτων ἐπ' εὐεργεcίᾳ τοῦ γένουc ἡμῶν κατελθόντι ποδαπὸν οἶκον ἄρα χρὴ καταcκευάζεcθαι; To contemplate and marvel at this φιλανθρωπία is to imitate the worship of the angels, who express their wonder at the manifestation of the divine δόξα on earth with the words πλήρηc ὁ οὐρανὸc καὶ ἡ γῆ τῆc δόξηc cου (Is. 6.3 and see Eusebius' comment on this text, cited above pp. 72f. on lines 47-48; on imitation of the angels in general see intro. above p. 6f.).

This understanding of φιλανθρωπία, however, is expanded in the lines that follow to include the grace imparted by Christ in his baptism and (by implication) the baptism of the faithful. Incarnation and baptism are closely linked in the following description of Christ's gracious descent, and the Greek and Coptic baptismal offices, in language similar to that of our hymn, speak of the source of this grace as τῆc cῆc ἀφάτου φιλανθρωπίαc and of God in this connection as φιλάνθρωποc (Goar, *Euchologion* 288; Burmester, "Baptismal Rite" 32).

**68-70 (30-31) κατῆλθεν πρὸc ἡμᾶc --- καὶ cαρκωθεὶc --- ἐκ τῆc ἀειπαρθένου καὶ βαπτιcθεὶc δι' ἡμᾶc:** ms. κατελθεν (η ⟩ ε Gignac I 242ff., but possibly due to neglect of temporal augment; cf. above p. 55 on ms. cυνεχθηcαν, 17-18 [dipl. 7]); ms. βαπτιcτηc (θ ⟩ τ frequent after c [Gignac I 87] here assisted by phonetic assimilation with previous -τιc-; ει ⟩ η Gignac I 240f.). These cola echo the language of creeds; e.g., *NCpol*: τὸν δι' ἡμᾶc τοὺc ἀνθρώπουc καὶ διὰ τὴν ἡμετέραν cωτηρίαν κατελθόντα ἐκ τῶν οὐρανῶν καὶ cαρκωθέντα ἐκ πνεύματοc ἁγίου καὶ Μαρίαc τῆc παρθένου καὶ ἐνανθρωπήcαντα (Lietzmann, *Symbolst.* 16/204 and cf. other creeds, ibid. 15f./203f.). The version of *NCpol.* preserved on tab. lignea Med.

inv. 71.00A (6th cent.; ed. O. Montevecchi, *Aegyptus* 55 [1977] 58-69) also uses ἀειπάρθενος, as does *Epiph. II* (Lietzmann, *Symbolst.* 15/203). On the word cf. Montevecchi, ibid. 66-69 and Horsely, *New Docs* 3 p. 116. Our text replaces ἐνανθρωπήσας with βαπτισθείς and subordinates it and σαρκωθείς to κατῆλθεν, as two aspects of Jesus' κατάβασις (on "descent," see also above lines 66-67 on ἴδετε κτλ. and lines 46-47 on ὁ κλίνας κτλ.).

With βαπτισθεὶς ὑπὲρ ἡμῶν cf. Just. *dial.* 88.4 (Migne PG 6.685b-c), where he says that Christ was baptized, οὐχ ὡς ἐνδεᾶ αὐτὸν τοῦ βαπτισθῆναι --- ὥσπερ οὐδὲ τὸ γεννηθῆναι αὐτὸν καὶ σταυρωθῆναι ὡς ἐνδεὴς τούτων ὑπέμεινεν, ἀλλ' ὑπὲρ τοῦ γένους τοῦ τῶν ἀνθρώπων. He continues to explain how Christ's baptism benefits the human race: it reveals Jesus as Messiah by the three-fold witness of John the Baptist, the descent of the Holy Spirit, and the voice from heaven proclaiming him Son of God.

This notwithstanding, the mentioning of the baptism of Jesus within a credal context gives pause, since standard versions of the *symbola* do not include it as a basic article of faith (Lietzmann, *Symbolst.* 10-20/198-208; J. Bornemann, *Die Taufe Christi durch Johannes* [Leipzig 1896] 55-58). Passages, however, from the second-century Ignatius, which have a credal flavor (see Lietzmann, *Symbolst.*, 44f./232f.), integrate the baptism of Jesus into other confessional affirmations; e.g., *Smyrn.* 1.1-2:[120]

εἰς τὸν κύριον ἡμῶν,
ἀληθῶς ὄντα ἐκ γένους Δαυὶδ κατὰ σάρκα,
υἱὸν θεοῦ κατὰ θέλημα καὶ δύναμιν θεοῦ,
γεγεννημένον ἀληθῶς ἐκ παρθένου,
βεβαπτισμένον ὑπὸ Ἰωάννου, ἵνα πληρωθῇ πᾶσα δικαιοσύνη ὑπ' αὐτοῦ·
ἀληθῶς ἐπὶ Ποντίου Πιλάτου καὶ Ἡρώδου τετράρχου καθηλωμένον κτλ.

[120]I reproduce here the colometry of Lietzmann, which is based on units of sense. The passage, however, does not seem to be rhythmical.

With this compare *Eph.* 18.2:[121]

| | Syllables | Stresses |
|---|---|---|
| Ὁ γὰρ θεὸϲ ἡμῶν Ἰηϲοῦϲ ὁ Χριϲτὸϲ | 12 | 3 |
| ἐκυοφορήθη ὑπὸ Μαρίαϲ | 11 | 2 |
| κατ' οἰκονομίαν θεοῦ | 8 | 2 |
| ἐκ ϲπέρματοϲ μὲν Δαυίδ, | 7 | 2 |
| πνεύματοϲ δὲ ἁγίου· | 7 | 2 |
| ὃϲ ἐγεννήθη καὶ ἐβαπτίϲθη, | 10 | 2 |
| ἵνα τῷ πάθει τὸ ὕδωρ καθαρίϲῃ. | 12 | 3 |

The second citation closely links Jesus' birth and baptism and designates the latter as an aspect of his redemptive πάθοϲ,[122] which brings us back to our hymnist's portrait of it as part of the divine *katabasis* .

Based on an *interpretatio christiana* of Biblical texts such as Ps. 73 (MT 74).13 (ϲὺ ἐκραταίωϲαϲ ἐν τῇ δυνάμει ϲου τὴν θάλαϲϲαν, ϲὺ ϲυνέτριψαϲ τὰϲ κεφαλὰϲ τῶν δρακόντων ἐπὶ τοῦ ὕδατοϲ), Patristic thought and baptismal liturgy conceive Jesus through his baptism as descending into the waters of chaos and subduing them and the dark forces within them.[123] By so doing, as Ignatius says above, he cleanses and sanctifies the baptismal waters for humans: Cyr. Hier. *cat. III de bapt.* 11, ἡγίαϲε τὸ βάπτιϲμα ὁ Ἰηϲοῦϲ βαπτιϲθεὶϲ αὐτόϲ. --- ὁ δράκων ἦν ἐν τοῖϲ ὕδαϲι κατὰ

[121]The colometry is mine. In contrast to the proceeding, this passage is a rhythmical poem of the same type as the "synagogue prayer" of *const. ap.* (see above p. 4). For ἡμῶν in the first line regarded as unaccented, see above on 14-15, p. 54.

[122]On this point and on this passage in general, see W.R. Schoedel, *A Commentary on the Letters of Ignatius of Antioch* (Hermeneia; Philadelphia 1985) 84ff.

[123]See Schoedel, loc. cit. (above n. 122). Christian iconography also exploited this theme; cf., e.g., the illustrated Khludov Psalter (9th cent.) at Ps. 74.13, which shows Jesus in the Jordan with John beside it, and on the banks a dismembered serpent (Schiller, I plate 359 with discussion and further material, p. 146f. [English trans. I p. 136f.]; G. Ristow, *Die Taufe Christi* [1965] plates 16 and 17 [pp. 32-33] and see the general thematic discussion , pp. 37-48 [esp. 45-48]).

τὸν Ιωβ ---. ἐπεὶ οὖν ἔδει cυντρίψαι τὰc κεφαλὰc τοῦ δράκοντοc, καταβὰc ἐν τοῖc ὕδαcιν ἔδηcε τὸν ἰcχυρὸν ἵνα ἐξουcίαν λάβωμεν πατεῖν ἐπάνω ὄφεων καὶ cκορπίων.

In the Coptic rite of the exorcism and blessing of the font, which immediately precedes baptism, are the following prayers: "Sanctify this water and this oil, that they may become a laver of regeneration. ... For it was he, thine only-begotten Son, our Lord Jesus Christ, who descended into the Jordan and purified it. ... Thou didst break the heads of the dragon upon the waters, ... O God the waters saw thee and were afraid" (Burmester, "Baptismal Rite" 39, 72-73; idem, *Coptic Church* 120-121). In general, cf. Bornemann, op. cit., pp. 58ff.; Daniélou, op. cit. (above p. 7 n. 15), 41f. In hymns, cf. Trypanis, "Three Hymns," 1.23ff., ὁ ἐν μεcονυκτίῳ,[124] φιλάνθρωπε, Ἰορδάνου τὰ ῥεῖθρα ἡγίαcαc, καὶ ἐμὲ --- τῶν πολλῶν ἁμαρτημάτων καθάριcον; similarly Ephraim Syrus, *hymn. de virg.* 15 (to the Jordan River), "Blessed are your torrents, cleansed by his descent. For the Holy One, who condescended to bathe in you, descended to open by his baptism the baptism for the pardoning of souls."[125] Our hymn thus portrays Jesus' baptism as an extension of the *katabasis* of his incarnation, and in this sense, as well those mentioned by Justin above, he is baptized ὑπὲρ ἡμῶν.

---

[124]On ἐν μεcονυκτίῳ see A. Dihle, *BZ* 69 (1976) 4f.

[125]The English translation is that of McVey (op. cit. above p. 27 n. 78) p. 326 and cf. n. 196. See the standard edition by E. Beck, *Des Heiligen Ephraem des Syrers Hymnen de Virginitate*, Corpus Scriptorum Christianorum Orientalium vol. 223 (Louvain 1962) pp. 51-54 (Syriac), and vol. 224 (Louvain 1962) 50-52 (German trans.) (Scriptores Syri vols. 94-95).

## Appendix: Summary of Phonological Features[126]

### Consonants (or involving consonants)

Stops

δ ⟩ τ: παρε**τ**ωθη 8/20; πλιαν**τ**α 25/59 (for Πλειάδα); ανεκ-**τ**ιεκετο (for -διήγητον) 29/67

τ ⟩ δ: **δ**ε 27/62

θ ⟩ τ: βαπτιc**τ**ηc [for -θείc] 31/70

γ ⟩ κ: **κ**αρ 9/23; ανεκτιε**κ**ετο (for -διήγητον) 29/67*

Nasals, liquids, spirants

om. final ν: εληιcω⟨**ν**⟩ ημαc 5/13; ο⟨**ν** ε⟩cταυρωcαν 8/21; ανεκτιεκετο⟨**ν**⟩ εαυτου 29/67; φιλανθρωπια⟨**ν**⟩ κατελθη 30/67

ν superfluous: πλιαντα (for Πλειάδα) 25/59[127]; παντω**ν** (for πάντα, see ad loc.) 11/30

om. final c: Ιηcου⟨**c**⟩ δια 3/4; Μαρια⟨**c**⟩ | ζωην 3/7; προφηται⟨**c**⟩ μεγαλοφωνοc 15/40

final c superfluous: θαλ{λ}αc⟨c⟩η{c} 28/64

final c ⟩ ν: Ιωαννη**ν** εβαπτιcεν 6/14*; δεcποτη**ν** πιρεν 9/23*

ν ⟩ ρ: περιπατω**ρ** ωc 23/57

---

[126]Diplomatic and edited text line numbers are separated by a slash. In the case of final letters I give also the next word (or word fragment) to show the phonetic context. References marked with * indicate interchanges possibly for reasons other than (or in addition to) phonetic; see commentary ad loc.

[127]Following the tendency of Byzantine-period papyri (Gignac I 114), movable -ν is used in all positions, before consonants (εἶδεν, 42; ὑμνοῦcιν, 43; κατῆλθεν, 68) as well as before vowels (ἔλαβεν, 24; τρι{α}cίν, 43).

μ ⟩ ν: cεραφιν υμνουcιν 17/43

Other

"false" or analogous aspiration: καθ ιδιαν 20/51

gemination and simplification: θαλλαcηc 24/57, 28/64

syllable loss: ⟨αμ⟩μων 28/64

**Vowels**

Diphthongs

αι ⟩ ε: cυμενει 2/2; γεννατε | Ιηcου⟨c⟩ 2/4

αι ⟩ υ: χυρωντεc 12/34*

ει ⟩ ε: αποκτινεν 5/11*

ει ⟩ ι: αποκτινεν 5/11; πιρεν (for πεῖραν) 9/24; cταυρωθιc 9/25; ιδεν 16/42; cαρκωθιc 21/53, 30/69; τεθιc (for τιθείc) 28/64

ει ⟩ η: βαπτιcτηc for -θείc 31/70*

οι ⟩ ε: ε (for οἱ) αγγελοι 12/34 (? see ad loc.); ε (for οἱ) πιcτη (i.e., -οι) 29/65

οι ⟩ η: πιcτη ιδετε 29/65

οι ⟩ υ: πυουν for ποιῶν 24/59

Simple Vowels

α ⟩ ε: πιρεν (for πεῖραν) 9/24; ραβδοcεντηc (for ῥαβδίcαντεc) 10/27

α ⟩ οι: ενδοξοι | δε 26/62*

α ⟩ ω: τω cεραφιν 17/43; παντων (for πάντα, see ad loc.) 11/30

ε ⟩ ει: εcπειρεν (for Ἕcπερον) 25/59

ε ⟩ η: ελ**η**ιcω(ν) for ἐλέηcον 5/13, 10/26, 14/39, 22/54; ραβ-δοcεντ**η**c (for ῥαβδίcαντεc) 10/27; cταυρωcαντ**η**c 10/28

⟨**ε**⟩cταυρωcαν 8/21*

⟨**ε**⟩ξεcια 27/62

η ⟩ ε: cυν**ε**χθηcαν 7/17*; ανεκτι**ε**κ**ε**το for -διήγητον 29/67*; κατελθεν 30/68*

η ⟩ ι: ελη**ι**cω(ν) for ἐλέηcον 5/13, 10/26, 14/39, 22/54; ξ**ι**ραc 24/57

η ⟩ υ: c**υ**μενει (for cημαίνει) 2/2

ι ⟩ ε: ανεξ**ε**χνιαcτα 26/61*; τ**ε**θιc (for τιθείc) 28/64

ι ⟩ ει: επ**ει** θαλλαcηc 24/57

ο ⟩ ε: εcπειρ**ε**ν (for Ἕcπερον) 25/59

ο ⟩ ω: εληιc**ω**(ν) for ἐλέηcον 5/13, 10/26, 14/39, 22/54; τ**ω**ν 6/15, 20/50, 22/56; παρετ**ω**θη 8/20; αυτ**ω**ν 10/27, 11/28; χυρ**ω**ντεc 12/34; ψαλλ**ω**ντεc, λεγ**ω**ντεc 13/36; ω for ὁ 21/53 (ω εκ), 22/56 (ω εκ-), 23/57 (ω περι-), 24/59 (ω πυ-), 26/61 (ω ποι-); αρκτουρ**ω**ν 25/60; ⟨αμ⟩μ**ω**ν 28/64

υ ⟩ ε: c**ε**νεcει 23/56

ω ⟩ ο: μαρτυρ**ο**ν 7/18

ω/ῳ ⟩ ου: ξυλ**ου** (for -ῳ) ὄν 8/21*; πυ**ου**ν (for ποιῶν) 24/59; ποι**ου**ν 26/61

οc ⟩ ου: αριθμ**ου** αγιοc 27/62*

## ADDENDA

p. 1 n. 2: See also H. Maehler, *The Oxford Classical Dictionary* (Oxford $1996^3$) 250B s.v.books, Greek and Roman.

p. 47 n. 101: The description of the "athbash" writing of the alphabet as "mystical" perhaps applies to certain contexts, but note that it was also a simple Greek school exercise; cf. E. Ziebarth, *Aus der antiken Schule* (Kleine Texte 65; Bonn $1913^2$) # 1; R. Cribiore, *Writing, Teachers, and Students in Graeco-Roman Egypt* (American Studies in Papyrology 36, Atlanta 1996) p. 39.

p. 53: For other relevant material from Justin Martyr's speculation on the birth of Jesus, particularly regarding Lk. 1.35, cf. *1 apol.* 46.5 (quoted below on addendum to p. 54); 32.14; 33.4; *dial.* 100.5; 105.1; and on these *1 apol.* passages, as well as those cited ad loc. above, see the most recent ed. (with notes) by M. Marcovich (Patristische Texte und Studien 38, Berlin/New York 1994) and the English trans. with full commentary by L. W. Barnard (Ancient Christian Writers 56, New York 1997). The most striking verbal parallel to our γεννᾶται Ἰηcοῦc διὰ λόγου θεοῦ, however, occurs not in a discussion of the incarnation, but of holy communion: οὐ γὰρ ὡc κοινὸν ἄρτον οὐδὲ κοινὸν πόμα ταῦτα λαμβάνομεν· ἀλλ' ὃν τρόπον διὰ λόγου θεοῦ cαρκοποιηθεὶc Ἰηcοῦc Χριcτὸc ὁ cωτὴρ ἡμῶν καὶ cάρκα καὶ αἷμα ὑπὲρ cωτηρίαc ἡμῶν ἔcχεν, οὕτωc καὶ τὴν δι' εὐχῆc λόγου τοῦ παρ' αὐτοῦ εὐχαριcτηθεῖcαν τροφήν, ἐξ ἧc αἷμα καὶ cάρκεc κατὰ μεταβολὴν τρέφονται ἡμῶν, ἐκείνου τοῦ cαρκοποιηθέντοc Ἰηcοῦ καὶ cάρκα καὶ αἷμα ἐδιδάχθημεν εἶναι (*1 apol.* 66.2). The meaning seems to be, that, as the logos effected its own transformation into flesh and blood in the historic incarnation of Jesus, so the logos in the eucharistic prayer effects its on-going "incarnation" in the bread and wine, so that both the presence of the flesh and blood of Jesus in history and their perpetual sacramental presence in the church are διὰ λόγου θεοῦ. The culminating transformation (μεταβολή) is that of believers, who are nurtured through the sacrament. For this passage and its difficulties, see Blunt, op. cit. (above p. 53) intro. (pp. xlff.) and ad loc. (pp. 98f.); Marcovich op. cit., p. 127; Barnard op. cit., p. 181; C. Munier, *L'apologie de Saint Justin philosophe et martyr* (Paradosis 38; Fribourg, Suisse 1994) 137f.

p. 54: With these four cola cf. esp. Just. *1 apol.* 46.5, διὰ δυνάμεως τοῦ λόγου κατὰ τὴν τοῦ πατρὸς πάντων καὶ δεσπότου θεοῦ βουλὴν διὰ παρθένου ἄνθρωπος ἀπεκυήθη καὶ Ἰησοῦς ἐπωνομάσθη. More specifically on "Jesus" as a designation of humanity, cf. τὸ δὲ Ἰησοῦς ὄνομα ⟨ἄνθρωπος⟩ τῇ Ἑβραΐδι φωνῇ, σωτὴρ τῇ Ἑλληνίδι διαλέκτῳ δηλοῖ (ibid. 33.7; ⟨ἄνθρωπος⟩ Marcovich); "Ἰησοῦς" δὲ καὶ ἀνθρώπου καὶ σωτῆρος ὄνομα καὶ σημασίαν ἔχει (*2 apol.* 6.4). Justin in fact derived Ἰησοῦς from Hebrew איש, "man" (see Marcovich on 33.7).

## Abbreviations and Select Bibliography


| | |
|---|---|
| *ACO* | *Acta conciliorum oecumenicorum*, 3 vols., ed. E. Schwartz (Berlin 1924-40) |
| Bauer/Aland | Bauer, W., Aland, K., Aland, B., *Griechisch-deutsches Wörterbuch zu den Schriften des Neuen Testaments und der frühchristlichen Literatur* (Berlin/New York[6] 1988) |
| BDR | Blass, F., Debrunner, A., Rehkopf, F., *Grammatik des neutestamentlichen Griechisch* (Göttingen[16] 1986) |
| Brightman | Brightman, F.E., *Liturgies Eastern and Western*, vol. 1, Eastern Liturgies (Oxford 1896) |
| Brock | Brock, S., "The thrice-holy hymn in the Liturgy," *Sobornost* 7.2 (1985) 24-34 |
| Burmester, "Baptismal Rite" | Burmester, O.H.E. KHS-, "The Baptismal Rite of the Coptic Church," *Bulletin de la Société d'Archéologie Copte* 11 (1945) 27-86 |
| Burmester, *Coptic Church* | Burmester, O.H.E. KHS-, *The Egyptian or Coptic Church* (Cairo 1967) |
| Charles | Charles, R.H., *A Critical and Exegetical Commentary on the Revelation of St. John*, 2 vols. (New York 1920) |
| CC | Corpus Christianorum |

*DACL* — *Dictionnaire d'Archéologie chrétienne et de Liturgie*, ed. F. Cabrol, H. Leclercq (Paris 1907-53)

Daniel/Maltomini *Suppl. Mag.* I, II — *Supplementum Magicum* vol. I, II ed. R.W. Daniel and F. Maltomini (Papyrologica Coloniensia XVI 1, 2), Abh. d. Rhein.-Westf. Akademie der Wissensch. (Opladen 1990, 1992)

Dornseiff, *Alphabet* — Dornseiff, F., *Das Alphabet in Mystik und Magie* (ΣΤΟΙΧΕΙΑ VII, Leipzig/Berlin 1925[2])

*EEC* I, II — *Enclyclopedia of the Early Church*, ed. A. Di Berardino, trans. A. Walford (New York 1992)

Engberding — Engberding, H., "Zum formgeschichtlichen Verständnis des ἅγιοϲ ὁ θεόϲ, ἅγιοϲ ἰϲχυρόϲ, ἅγιοϲ ἀθάνατοϲ, ἐλέηϲον ἡμᾶϲ," *Jahrbuch für Liturgiewissenschaft* 10 (1930) 168-74

Follieri — Follieri, E., *Initia hymnorum ecclesiae Graecae*, 5 vols. (Città del Vaticano 1960-66)

GCS — Die griechischen christlichen Schriftsteller der ersten drei Jahrhunderte (Leipzig 1897-1941; Berlin and Leipzig 1953; Berlin 1954ff.)

Gignac I, II — Gignac, F.T., *A Grammar of the Greek Papyri of the Roman and Byzantine Periods*, I Phonology, II Morphology (Testi e documenti per lo studio dell'antichità LV 1 and 2), Milano 1976 and 1981

| | |
|---|---|
| Goar, *Euchologion* | Goar, J., *Euchologion, sive rituale Graecorum* (Venice 1730²; repr. Graz 1960) |
| van Haelst | van Haelst, J., *Catalogue des Papyrus Littéraires Juifs et Chrétiens* (Paris 1976) |
| Hanssens | Hanssens, J.M., *Institutiones Liturgicae de Ritibus Orientalibus* III (Rome 1932) |
| Horsley, *New Docs* | Horsley, G.H.R., *New Documents Illustrating Early Christianity*, 1 (Macquarie 1981), 2 (1982), 3 (1983), 4 (1987) |
| KG | Kühner, R.; Gerth, B., *Ausführliche Grammatik der griechischen Sprache* II 1 and 2 (Hannover/Leipzig 1898, 1904) |
| Koenen | Koenen, L., "Ein Christlicher Prosahymnus des 4. Jahdt.s" in *Antidoron Martino David oblatum. Miscellanea papyrologica*, ed. E. Boswinkel, B.A. van Groningen, P.W. Pestman (Papyrologica Lugduno-Batava XVII, 1968) |
| Kotansky, *GMA* I | Kotansky, R., *Greek Magical Amulets*, Part I, Published Texts (Papyrologica Coloniensia XXII 1), Abh. d. Rhein.-Westf. Akademie der Wissensch. (Opladen 1994) |
| Lampe | Lampe, G.W.H., *A Patristic Greek Lexicon* (Oxford 1961) |
| Lietzmann, *Symbolst.* | Lietzmann, H., *Symbolstudien I-XIV*, Darmstadt 1966 (*Kleine Schriften* III, pp. 189-281; in citations pp. of the separate edition and of *KS* separated by slash) |
| Maas, *FK* | Maas, P., *Frühbyzantinische Kirchenpoesie*, Kleine Texte 52/53 (Berlin 1931²) |

| | |
|---|---|
| Maas/Trypanis | Maas, P., Trypanis, C.A., edd., *Sancti Romani Melodi Cantica; Cantica Genua* (Oxford 1963) |
| Maas/Trypanis *cant. dub.* | idem, *Cantica Dubia* (Berlin 1970) |
| MacCulloch | MacCulloch, J.A., *The Harrowing of Hell* (Edinburgh 1930) |
| Mayser | Mayser, E., *Grammatik der griechischen Papyri aus der Ptolemäerzeit* I 2-3 (2nd ed.), II 1-3 (Berlin/Leipzig 1926-38) |
| Mayser/Schmoll I 1 | Mayser I 1, bearb. von H. Schmoll (Berlin 1970[2]) |
| Migne, PG | Migne, J.P., Patrologia Graeca (Paris 1857-66) |
| idem PL | idem, Patrologia Latina (Paris 1844-55) |
| Mitsakis | Mitsakis, K., *The Language of Romanos the Melodist*, Byzantinisches Archiv 11 (München 1967) |
| Norden, *Ag. Th.* | Norden, E., *Agnostos Theos* (Darmstadt 1956) |
| *ODCC*[3] | *Oxford Dictionary of the Christian Church* (Oxford 1997[3]), ed. F.L. Cross, E. Livingstonc |
| *PGM* | *Papyri Graecae Magicae*, 2 vols., ed. K. Preisendanz, 2nd ed. with additions by A. Henrichs, Stuttgart 1973-74 (Christian charms in this collection are designated *PGM* P) |
| Pitra | Pitra, J.B., *Analecta Sacra Spicilegio Solesmensi* I (Paris 1876) |

| | |
|---|---|
| Psaltes | Psaltes, S.B., *Grammatik der Byzantinischen Chroniken* (Forschungen zur griechischen und lateinischen Grammatik 2), Göttingen 1913 |
| Quasten | Quasten, J., "Oriental Influence in the Gallican Liturgy," *Traditio* 1 (1943) 55-78 |
| Quecke, *Stundengebet* | Quecke, H., *Untersuchungen zum koptischen Stundengebet,* Publications de l'Institut Orientaliste de Louvain 3, 1970 |
| Rom. *cant.* | Romani Melodi *cantica*, cited from Maas/Trypanis |
| Rom. *cant. dub.* | Romani Melodi *cantica dubia*, cited from Maas/Trypanis *cant. dub.* |
| [Rom.] | works attributed to Romanus by various editors, but not in *cant.* or *cant. dub.* |
| SC | Sources Chrétiennes |
| Schermann | Schermann, Th., *Ägyptische Abendmahlsliturgien des erstens Jahrtausends,* Studien zur Geschichte und Kultur des Altertums 6 (Paderborn 1912) |
| Schiller | Schiller, G., *Ikonograpie der christlichen Kunst* I (Gerd Mohn 1966), II (1968), III (1971). English trans. (Greenwich Conn. I 1971; II 1972). |
| Schwyzer | Schwyzer, E., *Griechische Grammatik* I, München 1939, and II (with A. Debrunner) 1950 |
| *TDNT* I-X | *Theological Dictionary of the New Testament*, trans. G.W. Bromiley (Grand Rapids 1964ff.), Eng. trans. of *Theologisches Wörterbuch zum neuen Testament,* |

| | |
|---|---|
| | ed. G. Kittel, G. Friedrich (Stuttgart 1933ff.) |
| Teodorsson | Teodorsson, S.-T., *The Phonology of Ptolemaic Koine*, Studia Graeca et Latina Gothoburgensia 36 (Lund 1977) |
| Threatte, *Grammar* | Threatte, L., *The Grammar of Attic Inscriptions* (I Phonology, II Morphology; Berlin/New York 1980, 1996) |
| Trypanis "Three Hymns" | Trypanis, C.A., "Three New Early Byzantine Hymns," *BZ* 65 (1972) 334-38 |
| Turner, *GMAW* | Turner, E.G., *Greek Manuscripts of the Ancient World*, ed. P.J. Parsons, (London 1987[2]) |
| Werner | Werner, E., "The Genesis of the Liturgical Sanctus" in *Essays presented to Egon Wellesz*, ed. Jack Westrup (Oxford 1966) pp. 19-32 |
| Woolley, *Coptic Offices* | Woolley, R.M. (trans.), *Coptic Offices* (London 1930) |

# INDICES

## I. GREEK INDICES OF *P. MICH.* 799

Plain-faced numbers refer to the lines of the edited text with the corresponding lines of the diplomatic followed by a slash. Bold-faced numbers indicate the page of the introduction/commentary where the word or concept is discussed. II A/B means see or see also the appropriate section of index II. The letters *ns* by a line entry mean the word in that line is an abbreviated divine name (on *nomina sacra*, see above p. 30).

### A. Biblical Proper Names



### B. General Index of Words

## II. SUBJECT INDEX

(A) and (B) in a *see*/*see also* reference indicate the respective parts of this index.

### A. English

## **B. Greek** (see also I)

## III. SELECT INDEX LOCORUM

In this index M = Migne PG. For other abbreviations, see Abbreviations and Bibliography above or ad loc. Abbreviations of Classical and Patristic works for the most part follow the conventions of Liddell-Scott-Jones and Lampe.

### A. Biblical Literature

#### 1. LXX

2. NT



## B. Patristic Literature

1. GREEK

### 2. LATIN



## C. Liturgiae et Symbola



## D. Hymns

### *Adespota*

## E. Magical Texts

Thomas Gelzer, Michael Lurje, Christoph Schäublin

# Lamella Bernensis

## Ein spätantikes Goldamulett mit christlichem Exorzismus und verwandte Texte

1999. X, 196 Seiten mit 6 Abb. und einer Falttafel im Anhang. 15,5 × 23,5 cm
Geb. DM 86,– / ÖS 628,– / SFr 77,–
(BzA Band 124) ISBN 3-519-07673-X

Die ‚Lamella Bernensis' (LB) ist ein griechisches Amulett auf einer Goldfolie mit einem, von einem theologisch einigermaßen gebildeten Exorzisten verfaßten, in sieben Abschnitte gegliederten christlichen Exorzismus (wohl einem Tauf-Exorzismus) aus dem 5. Jh. n. Chr. für Leontios, Sohn der ‚heiligen Mutter' Nonna. Der Text steht in naher Beziehung zu dem Amulett für Alexandra, Tochter der Zoe, auf dem Silberstreifen der ‚Tablette Magique de Beyrouth' (TMB) im Louvre und zu einem Teil des Papyrus-Amuletts für Paulus Iulianus aus Oxyrhynchos (PSI 29 = PGM XXXV) in Florenz. Auch diese Texte wurden einer Nachprüfung unterzogen. Das Ergebnis der neuen Lesungen ist in synoptischer Darstellung auf einer Falttafel wiedergegeben. Daran sind Art und Verwendung gemeinsamer Vorlagen für unterschiedliche Zauberzwecke zu beobachten. Das vielfältige im Kommentar zur Erklärung herangezogene Material soll dazu dienen, LB und TMB (und die entsprechenden Teile von PGM XXXV) in den großen Zusammenhang der spätantiken und frühbyzantinischen Zauberpraxis und der magischen Tradition insgesamt einzuordnen. Aus der Diskussion ergeben sich auch Vorschläge zur Interpretation anderer verwandter Zaubertexte.

B. G. Teubner Stuttgart und Leipzig

## F. Other Papyri and Inscript.

(*see also*: Hymns, Magical Texts)



## G. Greek Literature